T0213672

Simula SpringerBriefs on Computing

Volume 10

Springer and Simula have launched a new book series, *Simula SpringerBriefs on Computing*, which aims to provide introductions to select research in computing. The series presents both a state-of-the-art disciplinary overview and raises essential critical questions in the field. Published by SpringerOpen, all *Simula SpringerBriefs on Computing* are open access, allowing for faster sharing and wider dissemination of knowledge.

Simula Research Laboratory is a leading Norwegian research organization which specializes in computing. The book series will provide introductory volumes on the main topics within Simula's expertise, including communications technology, software engineering and scientific computing.

By publishing the *Simula SpringerBriefs on Computing,* Simula Research Laboratory acts on its mandate of emphasizing research education. Books in this series are published only by invitation from a member of the editorial board.

More information about this series at https://link.springer.com/bookseries/13548

Kent-André Mardal · Marie E. Rognes ·
Travis B. Thompson · Lars Magnus Valnes

Mathematical Modeling of the Human Brain

From Magnetic Resonance Images to Finite
Element Simulation

Kent-André Mardal
Department of Mathematics
University of Oslo
Oslo, Norway

Travis B. Thompson
Mathematical Institute
University of Oxford
Oxford, UK

Marie E. Rognes
Department of Numerical Analysis
and Scientific Computing
Simula Research Laboratory
Oslo, Norway

Lars Magnus Valnes
Radiology and Nuclear Medicine
Oslo University Hospital
Oslo, Norway

ISSN 2512-1677   ISSN 2512-1685   (electronic)
Simula SpringerBriefs on Computing
ISBN 978-3-030-95135-1   ISBN 978-3-030-95136-8   (eBook)
https://doi.org/10.1007/978-3-030-95136-8

This Springer imprint is published by the registered company Springer Nature Switzerland AG
The registered company address is: Gewerbestrasse 11, 6330 Cham, Switzerland

Series Foreword

Dear reader,

Our aim with the series *Simula SpringerBriefs on Computing* is to provide compact introductions to selected fields of computing. Entering a new field of research can be quite demanding for graduate students, postdocs, and experienced researchers alike: the process often involves reading hundreds of papers, and the methods, results and notation styles used often vary considerably, which makes for a time-consuming and potentially frustrating experience. The briefs in this series are meant to ease the process by introducing and explaining important concepts and theories in a relatively narrow field, and by posing critical questions on the fundamentals of that field. A typical brief in this series should be around 100 pages and should be well suited as material for a research seminar in a well-defined and limited area of computing.

We have decided to publish all items in this series under the SpringerOpen framework, as this will allow authors to use the series to publish an initial version of their manuscript that could subsequently evolve into a full-scale book on a broader theme. Since the briefs are freely available online, the authors will not receive any direct income from the sales; however, remuneration is provided for every completed manuscript. Briefs are written on the basis of an invitation from a member of the editorial board. Suggestions for possible topics are most welcome and can be sent to aslak@simula.no.

January 2016

Prof. Aslak Tveito
CEO

Dr. Martin Peters
Executive Editor Mathematics
Springer Heidelberg, Germany

Foreword

Neuroscientists like to remind us that the brain is the most complex object in the known universe. The complexity they talk about is related to the brain's strange ability to integrate sensory inputs, to learn, to think, to store memories, to develop feeling, and to perform higher cognitive functions such as consciousness, self-awareness, mathematics, and yes, being able to write poems and equations about itself. Mostly, neuroscientists think about complexity in terms of signal processing and information transfer for which they have accumulated, through a century of exploration, an encyclopedic knowledge. Despite these Herculean efforts, much about the brain remains a mystery. In particular, there is another level of complexity associated with the brain that has been mostly neglected in the traditional neurosciences. The brain is a living organ that relies on a myriad of biological, chemical, and physical processes perfectly orchestrated to maintain its basic activities. Viewed from a physical perspective, what makes the brain so fascinating and so complicated as an organ is that it operates across multiple scales and constantly uses multiple physical fields. Indeed, processes that take place in the blink of an eye may be coupled with events that develop over a lifetime. In space, what happens inside a single neuron may trigger a global response at the organ or even body level. This large time and space scale-coupling prevents us from using the physical scale-separation paradigm that has been so successful in the study of planets and atoms. Similarly, the brain is a strange composite in which fluid and soft solid flow into one another, it is a soup of ions and electrolytes that needs to be carefully balanced at all times, and is a constant electric and magnetic field generator. This delicate symphony of processes is what allows the brain to function in harmony. Any defects or disturbance may lead to severe pathology as seen in development, trauma, or dementia. To make matter worse, our thick skull has impaired our ability to probe the brain properly and even some of its basic defining features, such as its material response under poking, are poorly understood.

Yet, for the last decade, many scientists from different fields have come together to rethink the way we think about the brain by adapting various theories and ideas from engineering, physics, mathematics, and computer science. To elevate brain modeling to a quantitative physical theory, one must combine data, experiments, theory, and computation. For many years, data was a true bottleneck as recording any physical fields in the living brain was particularly difficult and invasive. This situation changed completely with the advent of magnetic resonance imaging (MRI) that became routine in the late 1990s. The basic MRI and its multiple variants and generalizations have completely, but quietly, revolutionized medical practice by imparting a reliable, reasonably-high resolution, non-invasive means of observing the internal states of the brain. MRI imaging has become a basic source of information for clinical neuroscience and neurodegenerative disease research as it allows to map, and measure properties of the cortex and white matter, to determine the patterns of water flow within the brain, and to isolate regions of high cognitive processes.

In another realm of science, completely disconnected from neurosciences, another quiet revolution was taking place in the same period. With the rise of computing power, the ability to model and simulate the response of large three-dimensional structures through finite-element modeling also became routine. Initially, these methods were used to evaluate the safety of a bridge or to test the response of automotive pieces under loads. But, rapidly scientists realized that they could adapt these ideas to biology by simulating the response of arteries, heart, lungs, bones, and, eventually, the brain. The problem was not just to run existing codes to new soft structures but develop a new mathematical theory of soft biological materials. Indeed, with its extreme softness, viscous, active, nonlinear properties, and composite composition, the brain is not just a very soft piece of rubber but a complex material with fascinating properties not shared with any other organ.

Scientists interested in modeling the brain are now in an interesting situation: sitting on a giant heap of MRI data, with sophisticated theoretical and computational tools to simulate the brain, they have to find a way to bridge data to simulation and create a framework where systematic exploration of scientific and medical questions can be performed. This is where this little monograph comes in. This text, authored by four leading experts in the field, offers an explicit bridge linking MRI images to scientific computing and mathematical modeling of the brain. The authors introduce in simple and clear terms most of the concepts needed and provide a freely-available, open-source, and easy-to-use Python software tool allowing MRI images to be easily transformed into physiologically-accurate computational assets. They showcase their approach by showing how an anisotropic diffusion problem can be solved using a detailed computational domain, and diffusion tensor, constructed from a single patient MRI data set. Remarkably, what would have been a major research project a couple years ago can now be performed elegantly through their pipeline by any interested reader.

As Wittgenstein wrote in *Philosophical Investigations*: "We talk of processes and states, and leave their nature undecided. Sometime perhaps we will know more about them - we think. But that is just what commits us to a particular way of looking at the matter." Thanks to this wonderful book, now is the time when we will know more about processes and states of the brain.

Oxford Mathematical Institute   *Alain Goriely*
January, 2021

Preface

Observations surrounding the nature and fundamental biology of humankind date back to some of our earliest written historical accounts. Brain pulsatile behaviour and the structure of brain folding were described in the ancient Egyptian *Edwin Smith Surgical Papyrus* [1] dating back to 1700 BC. Hippocrates, the father of medicine, hypothesized that the brain was the *seat of intelligence*, while Aristotle was fascinated by and wrote about both *sleep* and *dreams*. However, early methods of directly investigating human anatomy were crude and invasive. Arguably, one of the profound medical achievements of our modern age is the advent of non-invasive imaging technologies.

The imaging revolution was born with Wilhelm Röntgen's discovery of X-rays as far back as in 1895. A few years later, Marie Curie successfully isolated radium, and X-rays then began to be used medically. Still, even with these early breakthroughs, positron emission tomography (PET) would not arrive until 1950, and it would be 1967 before the first clinical X-ray computed tomography (CT) scanner was put to use. An explosive development of different techniques followed these successes and the reader who is interested in the fascinating history of radiology can find more details in [65]. We now have many new imaging methods at our disposal; in addition to PET and CT, the various

---

[1] The text is named after the dealer who bought it.

magnetic resonance (MR) imaging techniques [2] have proven to be indispensable for understanding the brain of living patients. The medical imaging field is now vast. For instance, the annual meeting of the Radiological Society of North America hosts around 25,000 attendees, while 44,000 people attended the meeting of the Society for Neuroscience.

Meanwhile, engineering and industrial applications have led to the rapid development of both numerical methods, and applications using partial differential equations (PDEs) to model and understand physical phenomena. The finite element method (FEM), in particular, was introduced in the 1960s for solving PDEs on domains with complex geometries. Significant work has been invested in the construction of scalable, performant, and approachable software libraries for solving PDEs via the FEM. Today, we have many such libraries at our disposal; including Abaqus, COMSOL, deal.II, DUNE, and FEniCS. However, the generation of suitable, physiological finite element geometries for solving PDEs over brain domains remains a practical barrier. Therefore, the impact of computational modeling on medical imaging and neuroscience has not yet reached its full potential. Our aim with this book is to provide a bridge between common tools in medical imaging and neuroscience, and the numerical solution of PDEs that can arise in brain modeling. More specifically, our work focuses on the use of two existing tools, FreeSurfer and FEniCS, and one novel tool, the SVM-Tk, developed for this book.

A central, and practical, problem preventing a more widespread interest in the mathematical modeling of the human brain is that of anatomical mesh generation. Generating physiological finite element meshes of the brain is not an easy task. The sulcal and gyral folds of the cortex are intricate, and the extracellular diffusion tensor, dictated largely by axonal white-matter bundles, is anisotropic and tortuous. Nevertheless, such features are essential for even the simplest, patient-specific PDE models of brain structural deformation and fluid dynamics. The ability to accurately capture anatomical features could help us address many human problems, particularly when it comes to understanding the mechanisms underlying neurodegenerative pathology evolution. This book stands at the gateway of these pressing problems.

---

[2] MR images will play a fundamental role in this book. We introduce MR images in Chapter 2, but do take a sneak peek at Figures 2.1–2.3. It is astonishing that, in less than 100 years since Isidor Rabi published his seminal work measuring the nuclear spin of molecules [65], MRI has become a cornerstone of medical science and of mathematical modeling of the human brain.

Herein, we guide the reader through a straightforward methodology for ascertaining the basic assets, that is, a finite element mesh and the extracellular diffusion tensor, from a set of patient MR images. To do so, we introduce a novel software library, the Surface Volume Meshing Toolkit (SVM-Tk), wrapping functionality from the broad array of capabilities provided by the Computational Geometry Algorithms Library (CGAL), thus offering an approachable set of features specifically for the brain modeling community. Along the way, we will also employ the automatic segmentation capacity of the FreeSurfer tool set, a gold standard for MR image processing. The marriage of mathematical modeling, clinical imaging, and numerical analysis is demonstrated by solving a simplified PDE model of anisotropic gadobutrol diffusion in the brain.

We are deeply grateful to the numerous colleagues who have provided advice, and guidance, along the way as we commence our journey with you, the reader, into the exciting world of mathematical brain modeling. In particular, we thank Siri Fløgstad Svensson and Kyrre Eeg Emblem for the imaging data, Johannes Ring for creating Docker images, and Ana Budisa, Jørgen Dokken, Kyrre Eeg Emblem, Ingeborg Gjerde, Martin Hornkjøl, Miroslav Kuchta, Yngve Mardal Moe, Geir Ringstad, Vegard Vinje, and Bastian Zapf for their extremely constructive feedback on the book and the associated software. Jacob Schreiner has made significant contributions to SVM-Tk and finally, we wish to thank Anders M. Dale for hosting several research visits to La Jolla, which were instrumental for starting this project. Finally, this work has received funding from the European Research Council (ERC) under the European Union's Horizon 2020 research and innovation programme under grant agreement 714892 and the Research Council of Norway, grant 300305 and 301013.

Oslo, Norway   *Kent-André Mardal*
Oxford, United Kingdom   *Marie E. Rognes*
                                                                             Travis B. Thompson
Jan, 2021   *Lars Magnus Valnes*

Contents

Chapter 1
Introduction

Our brain is our most precious yet most mysterious organ. It consists of nearly 100 billion neurons, of which typically has 10,000 connections that extend up to a meter. As such, it is an intricate web that enable us to experience the world. In addition to neurons, the brain consists of about the same number of glial cells, around 700 kilometers of blood vessels, the extracellular matrix, and is surrounded by clear water-like cerebrospinal fluid, which together all work to maintain the delicate neurons' environment in a healthy state. At the whole-organ level, this is already incredibly complex, yet this is only part of the story; at any given time, various processes are happening in the brain, such as the electrical impulses between neurons and the complex chemical signaling that helps to maintain homeostasis. Due to the innate micro-scale complexity of the brain, a natural approach, in attempting to understand the brain's physiology and function, is offered by homogenized, continuum-based modeling; here, the focus is on modeling the large-scale behavior arising from the aggregate of small-scale contributions.

While continuum-based brain modeling has many applications, our motivation for writing this book and the corresponding software tools comes from recent theories concerning the restorative mechanisms of sleep. Recent theories consider the brain a porous and elastic (poroelastic) medium, where the elastic medium consists of the cells, and the fluid-filled pores are the extracellular matrix (of course, hyperelastic and viscoelastic materials [25, 15] could also be considered). In this setting, a paradigm shift was introduced by the glymphatic theory, proposed and developed over the last eight years by the Nedergaard group [33]. The glymphatic theory proposes that extra-cellular diffusion, as described in the seminal work of Syková and Nicholson [63], is

© The Author(s) 2022
K.-A. Mardal et al., *Mathematical Modeling of the Human Brain*,
Simula SpringerBriefs on Computing 10,
https://doi.org/10.1007/978-3-030-95136-8_1

not sufficient to explain the fundamental transport processes in the brain. In particular, pressure-driven convective flow is hypothesized to wash away the larger molecules of the metabolic waste products produced during the day [33, 36, 70]. Such metabolic waste proteins are observed to accumulate in patients with neurodegenerative diseases such as Alzheimer's or Parkinson's disease.

This topic has received a great deal of attention from the modeling community when it comes to the mechanisms at the microscopic level, e.g. [7, 19, 20, 31, 53, 58, 59]. However, very few works address the macroscopic level and how to employ modeling in a patient-specific manner, see e.g., [16, 43]. Mesh generation based on medical images, typically integrating images of different types, plays a critical role in achieving a patient-specific assessment. Our intention with this book is to equip the reader with software tools to perform studies of this kind.

Many other applications involve continuum-based models of the brain's physiology. For instance, alternative macroscopic theories involving the prion-related development of Alzheimer's disease have been proposed [22, 38]. Another interesting observation is that astronauts often experience visual impairments and are at risk of developing early dementia as a consequence of their periods in low or zero gravity. This seems to be a result of intracranial pressure changes and a shift in fluid volumes in intracranial compartments [10]. Another well-known computational modeling problem relates to epilepsy: specifically, the inverse problem of electroencephalography (EEG) in determining the source of an epileptic seizure via an elliptic PDE [27].

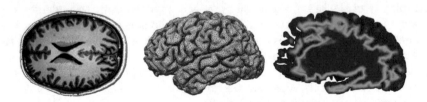

Fig. 1.1 Going from magnetic resonance (MR) images of a human brain to a numerical simulation of a biophysical phenomenon. From left to right: (a) An MR image of a human brain viewed along the axial direction, (b) a finite element mesh extracted from the MR image, (c) a snapshot of a tracer distribution simulation over this mesh. MR image types are discussed in Chapter 2.

This book provides the computational resources that form the foundations of continuum-based modeling of the human brain. Although we don't focus explicitly on the exciting multiphysics applications mentioned above, the approach discussed here is generalizable to multiphysics problems. You, the reader, will learn how to formulate, set-up and implement mathematical and computational models of brain biophysics in patient-specific geometries using finite element simulations and MR images (see Figure 1.1). We will use the evolution and distribution of a solute concentration due to diffusion as a model problem, and increase the complexity of the data and techniques involved over the course of the book. Of course, the process involves several challenges and pitfalls, which will be outlined.

1.1 A model problem

Suppose that we aim to study the diffusion of a solute concentration in a region of the brain. The region Ω could represent the left brain hemisphere or a smaller region such as the hippocampus, while the concentration u could represent an injected tracer used in imaging (such as gadobutrol [54] or dextran [34]) or possibly a metabolic waste protein, such as amyloid-β or tau. We can describe this model problem by a time-dependent partial differential equation (PDE): find the concentration $u = u(t, x)$ at spatial points $x \in \Omega$ and time points $t > 0$ such that

$$\begin{align}
u_t - \operatorname{div} D \nabla u &= f && \text{in } (0, T] \times \Omega, && \text{(1.1a)} \\
u &= u_d && \text{on } (0, T] \times \partial\Omega, && \text{(1.1b)} \\
u(0, \cdot) &= u_0 && \text{in } \Omega. && \text{(1.1c)}
\end{align}$$

In the diffusion equation (1.1a), u is the unknown field, while D is the diffusion tensor coefficient and f represents any source or sink for the concentration within the domain. The subscript t denotes the time derivative, div represents the divergence and ∇ the spatial gradient. The second equation (1.1b) gives a boundary condition: the function u_d represents a known distribution of the concentration on the boundary $\partial\Omega$ for all times. The third equation (1.1c) gives an initial condition for the solute concentration: the function u_0 represents the known initial concentration distribution throughout Ω at $t = 0$. The combined problem (1.1) is a complete initial boundary-value problem and will be our model problem.

1.2 On reading this book

This text does not assume that the reader is well versed in anatomy or neuroscience. In fact, most of the anatomical knowledge needed to follow along with this text is covered in Chapter 2.1. We have also made liberal use of footnotes and citations to inform the reader of additional interesting, or contextually useful, anatomical or physiological details. This text does, however, assume a basic knowledge of PDEs. For instance, the diffusion equation (1.1) is a classical continuum-based PDE with well-known behavior in both the mathematical and numerical sense. The reader unfamiliar with this equation is advised to first consult an introductory text on solving PDEs using the finite element method [24, 41, 42, 66].

The reader is assumed to be comfortable executing commands from a command line in a terminal window (also canonically referred to as a command window or command prompt). Terminal commands will be formatted throughout as:

```
$ cd ..
```

Commands at the operating-system (OS) level, such as that above, can differ from OS to OS, and we mainly demonstrate Linux commands here.

We also assume the reader is familiar with the fundamentals of the Python programming language or, alternatively, can understand the syntax well enough to follow the source code that appears throughout; we will not use any advanced Python programming techniques. We use Python 3 throughout, so please ensure that you have Python version 3.0 or any later version installed. You can check your Python version with either of the following terminal commands:

```
$ python --version
$ python3 --version
```

We will use the Python interface to the FEniCS Project finite element software [9], and we assume that the reader is familiar with the material covered by the FEniCS tutorial [41].

1.3 Datasets and scripts

The datasets and scripts used and described in this book are openly available and associated with its Zenodo community:
https://zenodo.org/communities/mri2fem/.

- The book dataset, including MR images, can be downloaded from
  http://doi.org/10.5281/zenodo.4386986 [45].
- The book scripts can be downloaded from
  http://doi.org/10.5281/zenodo.4386998 [46].
- A git repository containing the book and its scripts can be found at:
  https://github.com/kent-and/mri2fem.

We highly recommend that you download and unpack these materials before reading further. We expect to update the Zenodo community with script updates, updated installation guides and further material as needed.

1.4 Other software

We will use a number of external tools in this book. Most of these tools are available for several operating systems, with separate installation instructions and dependencies for each system. For the key external tools, we provide installation instructions for Linux Ubuntu (version 20.04, but earlier or later versions should also work along the same, or similar, lines). Whenever we refer to an Ubuntu-specific terminal command, we format it as follows:

```
$ sudo apt-get install ...                                    (Ubuntu)
```

We note that, before installing packages, it can be important to update the Ubuntu package list. This can be done by the following command:

```
$ sudo apt-get update                                         (Ubuntu)
```

For other operating systems, we refer to the specific software documentation for installation instructions.

1.5 Book outline

Chapter 2 provides an introduction to brain physiology and imaging, as and outlines the software ecosystem that will take us from MR images to numerical simulation. In Chapter 3, our aim is to get up and running quickly: we step through the entire pipeline from generating a volume mesh from MR image data to solving our model problem (1.1) on this mesh. In Chapter 4, we cover other aspects of meshing, including distinguishing between gray and white matter, merging left and right hemispheres, and adding parcellation labels. In Chapter 5, we focus on diffusion tensor imaging (DTI) and demonstrate how we can convert DTI data to a numerical tensor field. Finally, Chapter 6 brings together everything from Chapters 3 to 5 to present a realistic simulation of anisotropic diffusion in heterogeneous brain regions.

Chapter 2
Working with magnetic resonance images of the brain

2.1 Human brain anatomy

The human brain consists of multiple structures, including the large cerebrum, the smaller cerebellum, and the brain stem. These structures can easily be identified from MR images of the brain (Figure 2.1). The cerebrum is composed of the left and right hemispheres, which are connected through bundles of nerve fibers (the corpus callosum).

There are two main types of brain cells: neurons (or nerve cells) and glial cells. Each neuron is generally composed of a cell body (*soma*), a long nerve fiber (*axon*), and other branching extensions (*dendrites*). The spatial distribution of the brain cells results in two primary types of brain tissue matter: white and gray matter.[1] White matter mainly consists of bundles of (myelinated) axons while gray matter includes neuronal cell bodies, and glial cells. The distribution of white and gray matter in the cerebrum is shown in Figure 2.1 (right), which demonstrates the white matter and the cortical and sub-cortical gray matter. The sub-cortical gray matter includes a number of

---

[1] In simple terms, neurons have a cell body (soma) and branches (axons and dendrites) that extend from the cell body. Axons are either covered with a lipid-rich (fatty) layer called myelin (myelinated axons) or surrounded by other cells (unmyelinated axons); myelin helps the axons conduct electrical signals over long distances. Myelin, being fatty, gives off a white-ish color; conversely, unmyelinated axons, dendrites, and neural cell bodies, in close proximity to microcirculation, appear gray. This is the origin of the terms *gray matter* and *white matter*.

K.-A. Mardal et al., *Mathematical Modeling of the Human Brain*,
Simula SpringerBriefs on Computing 10,
https://doi.org/10.1007/978-3-030-95136-8_2

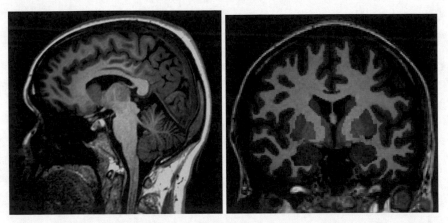

Fig. 2.1 MR images (T1-weighted) of the human brain. Left: A sagittal (longitudinal) cross-section of the human brain with the cerebrum in blue, the cerebellum in red, and the brain stem in yellow. Right: A coronal slice of the brain. The tissue matter composition of the cerebrum is marked by color with red denoting the cortical gray matter, blue denoting the subcortical gray matter and white denoting the white matter. (The colors were added after post-processing.)

important structures or regions located deep inside the brain, such as the hippocampus, basal ganglia and thalamus.

The brain is enclosed and protected by three layers of meninges; the outermost *dura*, the middle *arachnoid* and the innermost *pia* membrane. The narrow space between the pial and arachnoid membranes is filled with cerebrospinal fluid (CSF) and is referred to as the *subarachnoid space* (SAS). The SAS extends around the brain and further down along the spinal cord. The SAS is also connected to the ventricular system, a system of interconnected CSF compartments surrounded by the brain. The ventricular system consists of the two lateral ventricles and the third and fourth ventricles, shown in Figure 2.2. The thin passage between the third and fourth ventricle is known as the cerebral aqueduct.

We refer the reader, for example, to [26] for a good introduction to human brain anatomy and [13] for a general introduction to human physiology.

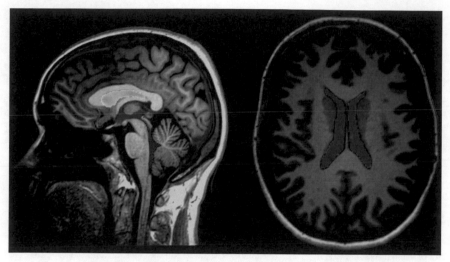

Fig. 2.2 Sagittal (left) and axial (right) MR image cross-sections of the brain with the lateral ventricles marked in yellow, the third ventricle marked in blue and the 4th ventricle and the aqueduct marked in red. The white region surrounding the lateral ventricles, on the left, is the corpus callosum. The distinction between (cortical) gray and white matter is also clearly visible on the right.

2.2 Magnetic resonance imaging

Magnetic resonance imaging (MRI) is a rich and versatile technique for the non-invasive medical imaging of the brain and other organs. The method has its roots in studies of Isidor Rabi; dating to the early 20th century. Rabi was able to ascertain information on the rotation and magnetic movements of the nuclei of atoms and molecules [65]. MRI leverages these types of magnetic properties; specifically for the nucleus of elemental hydrogen, which is abundant in fat and other tissues [12]. An MRI scanner creates a strong magnetic field, aligning the poles of hydrogen atoms along the scanner's axis. A radio wave is added to the magnetic field, causing the hydrogen nuclei to resonate, and different (scanner) slices of the body resonate differently. Turning off the applied radio frequency induces the realignment of the hydrogen nuclei with the applied magnetic field, and this process in turn causes the emission of another radio wave signal which the MRI scanner measures. The intensity of this last signal is what we visualize as MRI images [12].

The MRI method can be used for detailed investigations of tissue morphology and structure (structural MRI), tissue properties (e.g. diffusion-weighted MRI), blood flow (perfusion MRI) as well as aspects relating to brain function (functional MRI). As detailed above, an MR image is constructed from the interaction between a strong magnetic field and radio waves. The specifics of the procedure can be controlled by manipulating factors that affect this interaction, such as the magnetic field gradient. A specific set of changing magnetic gradients is referred to as an MRI *sequence*, which in turn has a number of parameters. We briefly describe a few key MRI sequences here; the interested reader can find more information in e.g. [29, 50, 8].

2.2.1 Structural MRI: T1- and T2-weighted images

Structural MRI provides high-resolution images of brain anatomy and can thus give information about the shape, size and composition of different brain compartments and regions[2]. Examples of structural MRI sequences include T1- and T2-weighted images, as already encountered in Figures 2.1 and 2.2. Such images are well suited for defining brain geometry models and will be used extensively for this purpose (Chapters 3–4). A brief introduction to T1- and T2-weighted images can be found in [52].

T1- and T2-weighted images correspond to different (groups of) MRI sequences, each with their own parameters and characteristics. A T1- or T2-weighted image is a three-dimensional image, typically viewed as a stack of black and white images of different planes (axial, coronal or sagittal) of the brain. The colored shading is referred to as the *signal intensity*; white represents a high signal intensity and black represents low intensity (with gray tones representing intermediate values). Both imaging sequences (T1- and T2-weighted) produce different signal intensities for different types of tissue and fluids, but the two differ in their dominant intensities.

In T1-weighted images, fat gives off a high intensity signal and appears white, while fluids give off a low intensity signal and appear black. Therefore, the brain tissue appears as different shades of gray with white matter appearing lighter than gray matter (Figure 2.3 (left)). T1-weighted images are effective at differentiating between white and gray matter, but less effective

---

[2] Gray matter (cortical) and white matter regions are defined vis-a-vis a segmentation and parcellation process. Parcellations are discussed in more detail in Chapter 4.2.1

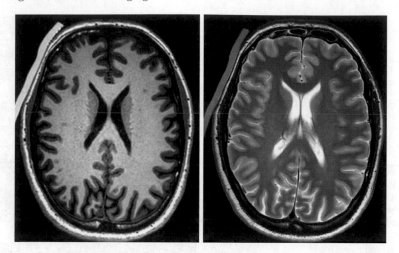

Fig. 2.3 T1-weighted image (left) versus T2-weighted image (right). In the T1-weighted image, white matter is recognized as light gray, whereas the darker gray lining the surface of the brain is gray matter. T1-weighted images are used particularly because they exhibit a sharp contrast between gray and white matter. The T2-weighted image shows the CSF as almost white and provides good contrast between the CSF and the brain, but less contrast between white and gray matter. Also note that blood is dark in T2-weighted images.

at distinguishing between, for example, the CSF (black) and the gray matter (dark gray). In particular, it can be difficult to identify fluid compartments such as the ventricles, the aqueduct, and SAS from T1-weighted images alone.

In T2-weighted images on the other hand, fluids give off a high intensity signal and appear white (Figure 2.3 (right)). Such images are less effective at distinguishing between white and gray matter, but can provide good contrast between the CSF (white) and brain matter (gray). T2-weighted images can thus supplement data from T1-weighted images for identifying and separating the ventricles and aqueduct from subcortical gray matter.

2.2.2 Diffusion-weighted imaging and diffusion tensor imaging

Diffusion MRI (dMRI) is an imaging modality that detects water molecule movement patterns [37, 60]. Both isotropic and anisotropic diffusion coeffi-

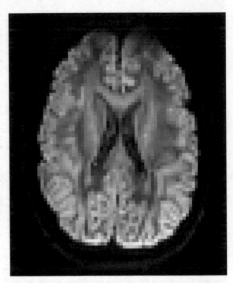

Fig. 2.4 A raw, axially-oriented dMRI image of the brain. The ventricles (center) appear in lower contrast due to the fast signal decay caused by the strongly isotropic diffusion of the water contained within them.

cients can be determined. Diffusion tensor imaging (DTI) is a specific type of diffusion-weighted MRI; Figure 2.4 shows an example of a DTI-weighted image. At a high level, a reference signal is used as a comparison and the DTI imaging process measures the difference in that reference signal with several follow-up signals. The resulting sequence provides information about how water diffuses in different regions of the brain.

More specifically, DTI provides information on both the magnitude and (multiple) directions of the movement of molecular water in the brain; that is, how water travels. In mathematical terms, DTI provides information regarding the diffusion tensor, D, in (1.1). In three dimensions, this tensor has nine entries and a global representation given by

$$D = \begin{pmatrix} d_{11} & d_{12} & d_{13} \\ d_{21} & d_{22} & d_{23} \\ d_{31} & d_{32} & d_{33} \end{pmatrix}. \tag{2.1}$$

Each of the entries d_{ij} can be a function of the position $x \in \mathbb{R}^3$. We will extract the heterogeneous, anisotropic, and patient-specific diffusion tensor from DTI image data in Chapter 5. DTI has been used extensively to study the layout of brain's white matter tracts; these tracts heavily bias the movement of water within the brain. We refer the interested reader to [37], and the many sources therein, for a discussion of the more advanced topics related to the field of dMRI techniques.

2.3 Viewing and working with MRI datasets

2.3.1 The DICOM file format

Medical images, including T1, T2 and DTI, are often stored in the DICOM file format. DICOM stands for *Digital Imaging and Communications in Medicine* and is an imaging standard [48]. The format stores both the image itself and a comprehensive set of meta-data, such as the imaging protocol and patient identification, which enables consistent and safe usage across different vendors and software packages.

The DICOM format gives the output from an MRI scan as a collection of files arranged in sequences. A given file within a sequence [3] also contains the necessary information to inform the viewer what the next (or previous) file in that file sequence is. DICOM files are sometimes stored in a binary file named DICOMDIR that indexes the entire structure of an MRI dataset.

2.3.2 Working with the contents of an MRI dataset

A DICOM viewer is an essential component for viewing and working with DICOM files. A number of DICOM browsers exist, and we will use Dicom-Browser [11]. Currently, DicomBrowser is available [4] as a pre-packaged binary download for several operating systems; the source code [5] is also available. On

---

[3] Formally, the term *MRI sequence* does not refer to an ordered measure (such as a sequence in time) but rather to a specific type of pulse sequence or field gradient that determines the specific imaging protocol. A T1 *MRI sequence* produces T1 images, a T2 *MRI sequence* produces T2 images, and so forth. Here, when we speak of 'a sequence of files' we are referring to the order in which the files are produced by the MRI scanner; this is typically reflected in the naming of the files such that IM_0001 would be produced before IM_0002, and so on.

[4] https://wiki.xnat.org/xnat-tools/dicombrowser

[5] https://bitbucket.org/xnatdcm/dicombrowser/src/master/

Ubuntu Linux, DicomBrowser comes pre-packaged as a Debian package file; this type of file can be installed on Ubuntu using the command

```
$ sudo apt install /path to file/my-dicombrowser.deb    (Ubuntu)
```

where you should change 'path to file' to the path where the DicomBrowser installation file (ending in .deb) has been downloaded and change the text of 'my-dicombrowser.deb' to be the specific filename of the DicomBrowser installer package.

You can then begin the process of extracting sequences from an MRI dataset by launching the DicomBrowser utility with the command:

```
$ DicomBrowser &
```

After the DicomBrowser window opens, select ⌈File→Open⌋ from the top menu bar; a dialogue box will appear with the heading ⌈Select DICOM files⌋. Navigate to the directory containing the (example) book dataset and select the directory titled dicom/ernie. In this directory, select the file titled DICOMDIR. The main DicomBrowser window should now show a list of the patients whose data are included in this dataset (in this case just one patient). In the main window (c.f. Figure 2.5 (left)), click the patient, whose ID starts with 1.3.46, and then click the symbol to the left of the patient ID to expand the entry.

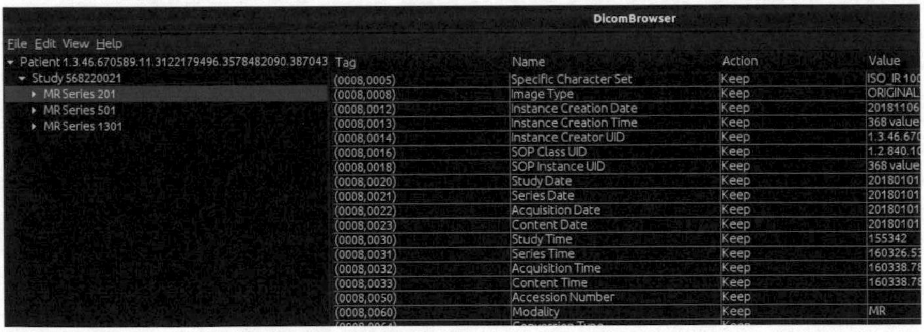

Fig. 2.5 Example of the DicomBrowser layout and selection of an MR series (dicom/ernie).

A list of studies now appears under the patient ID; generally, a patient can have several associated studies but the (example) book dataset contains one study per patient. Expand the study (beginning with 568) by clicking the symbol to the left. A list of MR series now fills the main window; many MR series can be contained in a single study and we see that the `dicom/ernie` data contains three series (201, 501 and 1301), as shown in Figure 2.5.

Find MR Series 201 and left-click on it. The secondary window now contains a list of tags, each of which has an associated name, action, and value (Figure 2.5). Note that tag (0018, 1030) [6] is named 'protocol name' and has value T13D; this indicates that MR Series 201 corresponds to a T1-weighted image sequence. With 'MR Series 201' as the current selection, in the DicomBrowser (left) window, we can select ⌈File→Save⌉ to extract the series; a window then appears which provides extraction options. We would like to extract this series to a directory with a memorable name; we can do this by changing '-anon' to '-T13D' to signify that the series we are extracting represents a T1-weighted image in 3D. Also note that DicomBrowser can be used to anonymize the data, as indicated by the default extension -anon, or, in general, to change DICOM tag values either through the GUI or in scripts.

We can use the same procedure to extract other image sequences. In the DICOM images in our example dataset, MR Series 501 is a DTI sequence and MR Series 1301 includes T2-weighted images. We have already extracted all three series present in the sample `DICOMDIR` into the (example) book dataset. Image series 201 has been extracted to `dicom/ernie/T13D`, image series 1301 has been extracted to `dicom/ernie/T23D`, and image series 501 has been extracted to `dicom/ernie/DTI`. We note that `dicom/ernie/DTI` also contains some non-image files; these files are not in the `DICOMDIR` data but will be generated and discussed in Chapter 5.

2.4 From images to simulation: A software ecosystem

In this section, we provide a brief overview of the software tools that comprise the pipeline used in this book. We will extract data from clinical images, segment the data, extract and work with diffusion tensor information, generate a finite element mesh, and bring all of this together to solve a partial differential equation. Along the way, we will need additional tools for miscellaneous

---

[6] Tag identifiers are not standard, and can differ based on the MRI scanner manufacturer. Generally, we identify a sequence from the Name field by looking for a label containing T1, T2, or DTI.

objectives, such as file conversion and visualization. A visual representation of the patient-specific pipeline from MRI-images to finite element simulations is shown in Figure 2.6.

2.4.1 FreeSurfer for MRI processing and segmentation

FreeSurfer [18] is an open-source software suite for the segmentation (identification of different brain regions), processing, visualization, and analysis of human brain MR images. FreeSurfer is well established, well documented, and widely used and we refer to the FreeSurfer website for extensive documentation, online tutorials, support, an overview of publications and general installation instructions [2]. Generally, all FreeSurfer commands also have extensive documentation available via --help. At the time of writing, FreeSurfer provides step-by-step installation guides for Linux [7] and Mac [8] Below, we discuss the Linux installation process, for the sake of completeness.

To install FreeSurfer on Ubuntu Linux, we suggest downloading the most recent FreeSurfer tar archive [9] locally, for instance under /home/me/local/src. If the name of this file is freesurfer.tar.gz, unpack the archive via

```
$ tar -zxvpf freesurfer.tar.gz                          (Ubuntu)
```

The next step is to configure your environment for using FreeSurfer. If your FreeSurfer archive has been unpacked at /home/me/freesurfer, you can configure your FreeSurfer environment manually by adding the following lines at the end of the file named .bashrc in your home directory:

```
export  FREESURFER_HOME=/home/me/freesurfer
export  SUBJECTS_DIR=$FREESURFER_HOME/subjects
source  $FREESURFER_HOME/SetUpFreeSurfer.sh
```

The FreeSurfer programming team requires that each user acquire a license file to use the software; the license file is free, and to acquire it, follow the registration directions at the FreeSurfer website [2]. Finally, FreeSurfer comes pack-

---

[7] https://surfer.nmr.mgh.harvard.edu/fswiki//FS7_linux

[8] https://surfer.nmr.mgh.harvard.edu/fswiki//FS7_mac

[9] See for example https://surfer.nmr.mgh.harvard.edu/fswiki/FS7_linux

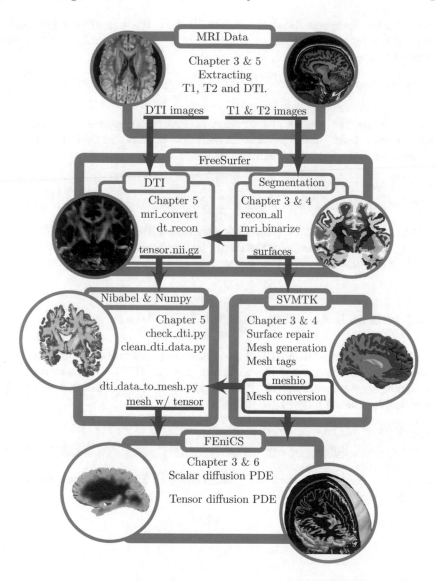

Fig. 2.6 Overview of the imaging, segmentation, meshing, simulation and visualization pipeline. T1, T2, and DTI images have already been discussed. Freesurfer and segmentation are discussed in Section 2.4.1. Nibabel and Numpy are Python libraries that are used for neuroimaging applications and convenient manipulation of tensors, respectively. SVM-Tk is a computational geometry package, written for this book, which is specialized for creating brain meshes and FEniCS is a Python library for high-performance finite element method computations.

aged with the Freeview visualization tool. Test Freeview (and your FreeSurfer installation) by opening a new terminal (noting the FreeSurfer environment commands appearing on top), and typing:

```
$ freeview &
```

to open the Freeview user interface.

We will use the command dt_recon in FreeSurfer, which has additional requirements: tcsh[10] and FSL. The installation of tcsh can be done from the terminal with the following command lines:

```
$ sudo apt update
$ sudo apt install tcsh                                        (Ubuntu)
```

FSL is a comprehensive library for functional MRI, MRI, and DTI brain imaging data [35], and we refer the reader to its website for installation instructions.[11] Note that FSL, like FreeSurfer, requires a license to be operational.

2.4.2 NiBabel: A python tool for MRI data

The Python module NiBabel [14] provides read and write access to several neuroimaging file formats. The module is part of NIPY, [12] a community for neuroimaging data analysis via Python. NiBabel can be installed using pip, for example, via the following terminal command:

```
$ sudo apt install python3-pip
$ sudo pip3 install nibabel                                   (Ubuntu)
```

We will use Nibabel to work with image data in Python in Chapter 4 and onwards.

---

[10] tcsh refers to a specific type of Unix shell. A Unix shell is a command-line interpreter; many such shells can be used in a Unix environment (such as Ubuntu Linux).

[11] See https://fsl.fmrib.ox.ac.uk/fsl/fslwiki/FslInstallation.

[12] See https://nipy.org/.

2.4.3 SVM-Tk for volume mesh generation

The Surface Volume Meshing Toolkit (SVM-Tk) generates meshes with sub-domains based on surfaces and segmentations provided by FreeSurfer. In particular, SVM-Tk provides a Python-based interface to the Computational Geometry Algorithms Library (CGAL) [21] and provides tools for both fixing and marking surfaces to enable a relatively robust mesh generation for the various compartments of the brain. We refer to the SVM-Tk documentation[13] for general installation instructions, including a detailed list of dependencies (including CGAL and a number of packages available, e.g. through the Ubuntu package manager).

2.4.4 The FEniCS Project for finite element simulation

We will use the open-source FEniCS Project [9, 44] as our finite element software. FEniCS includes both a C++ and a Python interface; we will use the Python interface throughout. Extensive documentation, support, an overview of publications, and general installation instructions can be found on the FEniCS Project website [1]. In particular, we strongly encourage the reader to become familiar with FEniCS and the finite element method via the introductory FEniCS tutorial [41].

There are many ways to install the FEniCS Project software, including the use of Docker images, using pre-built Anaconda packages, or from source. Two simple ways of installing FEniCS are via Docker images (see the FEniCS Project website[14]) and via Ubuntu package managers. For the latter, use the following terminal commands:

```
$ sudo apt-get install software-properties-common
$ sudo add-apt-repository ppa:fenics-packages/fenics
$ sudo apt-get update
$ sudo apt-get install --no-install-recommends fenics
```

---

[13] See https://github.com/SVMTK/SVMTK.

[14] https://fenicsproject.org/download/

2.4.5 ParaView and other visualization tools

We recommend using ParaView [6] to visualize the simulation results and other meshing objects that we discuss throughout the book. For installation instructions, see the ParaView website[15]. On Ubuntu Linux, ParaView can be installed with the following terminal command:

```
$ sudo apt-get install paraview                          (Ubuntu)
```

After installation, ParaView can be launched with

```
$ paraview &
```

An optional but useful tool for visualizing surfaces (in the form of surface STL files) is Gmsh [23]. For installation instructions, see the Gmsh website.[16] On Ubuntu Linux, Gmsh can be installed with the following terminal command:

```
$ sudo apt-get install gmsh                              (Ubuntu)
```

For quick plotting in Python, the Python package Matplotlib is very convenient [32]. The pyplot module of Matplotlib can be imported in a Python script (as any other Python module) as:

```
import matplotlib.pyplot as plt
```

2.4.6 Meshio for data and mesh conversion

We recommend using meshio [56] for conversion between computational mesh formats. The meshio suite can convert between various unstructured mesh formats, for instance, between the CGAL medit file format (.mesh) and the FEniCS mesh format (.xml or .h5). We suggest using the pip installer to install meshio[17] for example, via typing the following in the terminal:

---

[15] See https://www.paraview.org/.

[16] See http://gmsh.info/#Download.

[17] See https://pypi.org/project/meshio/.

```
$ sudo pip3 install meshio[all]                           (Ubuntu)
```

2.4.7 Testing the software pipeline

To verify that all software dependencies are correctly installed, we provide a test script (in **mri2fem/chp2**); the test script can be executed with the command:

```
$ python3 test_book_software.py
```

The script will check each software dependency and output a response indicating whether it is installed or not. If the software is not installed, the response provides a detailed description of ways to correctly install it, including links to installation guides:

```
Checking FEniCS installation.

    FEniCS is not installed
    Follow the installation guide at
    https://fenicsproject.org/
```

If the software is installed, the response is a single-line response confirming that the software is installed:

```
Checking FEniCS installation. => FEniCS installed
```

Chapter 3
Getting started: from T1 images to simulation

The goal of this chapter is to outline how to perform a numerical simulation of a brain region defined from structural MR images. To address this challenge, we first demonstrate how to generate a high quality mesh of a brain hemisphere from T1-weighted MR images using the tools introduced in Chapter 2. Next, we show how to define a finite element discretization of the diffusion equation (1.1) over this mesh to simulate the influx of an injected tracer. [1]

3.1 Generating a volume mesh from T1-weighted MRI

To generate a mesh from an MRI dataset including T1-weighted images, we follow three main steps:

1. Extract a T1-weighted image series from the MRI dataset,
2. Create (boundary) surfaces from the T1-weighted images using FreeSurfer,
3. Generate a volume mesh of the interior using FreeSurfer along with SVM-Tk.

---

[1] MRI scanners operate by manipulating a magnetic field around the patient and then measuring the body's molecular interaction with radio waves. Non-toxic substances, designed specifically with this interaction in mind, have been developed that can be injected into a patient to improve the visibility of internal body structures. These substances are referred to as *MRI tracers* or *MRI contrast agents*. Dennis Carr and Wolfgang Schorner published the first tracer-enhanced MR images, in humans, using gadolinium diethylenetriaminepentacetate [65].

© The Author(s) 2022
K.-A. Mardal et al., *Mathematical Modeling of the Human Brain*,
Simula SpringerBriefs on Computing 10,
https://doi.org/10.1007/978-3-030-95136-8_3

We will consider each of these steps in order, and encourage the reader to ensure that FreeSurfer is installed and configured (see Chapter 2.4.1) before proceeding. Step 2 can be particularly time consuming, probably involving FreeSurfer segmentation and reconstruction run times of up to 24 hours. While we provide already processed files in the tarball that accompanies the book, we encourage the reader to try these steps.

3.1.1 Extracting a single series from an MRI dataset

To extract a single MR series from an MRI dataset or a DICOM database, we can use the DicomBrowser graphical interface or FreeSurfer command line tools. The first option, using the DicomBrowser graphical interface to extract a T1-weighted image series, is described in Chapter 2.3.2 with the book dataset as an example, and the resulting image series can be found under dicom/ernie/T13D in the book dataset. The second option is described in Chapter 3.4.

3.1.2 Creating surfaces from T1-weighted MRI

The next step is to create surfaces, representing, for example, the interface between the pial membrane and the surrounding cerebrospinal fluid (CSF), referred to here as the pial surface (Figure 3.1), or the interface between white and gray matter, from the T1-weighted MRI series just extracted. We will use FreeSurfer for this task, and as an example, we will extract the pial surface surrounding the left hemisphere of a brain.

To conduct a full image stack segmentation and surface reconstruction, we take advantage of the FreeSurfer command `recon-all`. This command is compute-intensive, with likely run times of up to 24 hours. To invoke `recon-all`, we select one of the T1 DICOM files: e.g., in `dicom/ernie/T13D`, we can pick `IM_0162`. Next, we decide on a subject identifier for the FreeSurfer pipeline; we choose to name this example subject "ernie". We are now ready to launch `recon-all`: [2]

---

[2] It is advisable to do this on a separate core or overnight.

```
$ cd dicom/ernie/T13D
$ recon-all -subjid ernie -i IM_0162 -all
```

The results of `recon-all` will be output to the folder SUBJECTS_DIR/ernie, where the environment variable SUBJECTS_DIR defaults to the `subjects` folder under FREESURFER_HOME (see Chapter 2.4.1). For convenience, this output is also included in the book dataset, under `freesurfer/ernie`. If we inspect the contents of this directory we will see several subdirectories. Some important subdirectories are:

- `/stats`, contains files providing statistics derived during segmentation;
- `/mri`, contains volume files generated during segmentation; and
- `/surf`, contains surface files generated during segmentation.

To view the generated surface files, we focus on the `/surf` directory, and launch Freeview (see Chapter 2.4.1):

```
$ cd freesurfer/ernie/surf
$ freeview &
```

Targeting the pial surface that surrounds the left brain hemisphere as an example, select File→Load Surface from the command bar, and then select the file titled `lh.pial` (where `lh` refers to left hemisphere and `pial` denotes the pial surface). After a moment, the view windows will be populated with 2D surface slices shown as curves, in addition to a reconstructed 3D image of the pial surface (Figure 3.1).

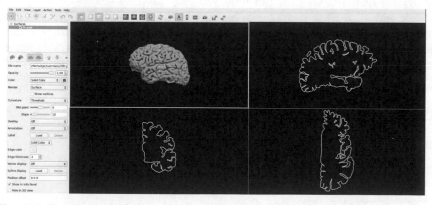

Fig. 3.1 Freeview visualization of the pial surface of a single brain hemisphere extracted from T1 images via FreeSurfer.

To work further with this surface, we use the FreeSurfer tool `mris_convert` to convert the binary surface file into the STL format [55]. STL is a widely used format representing the surface discretely in terms of vertices, triangles and normals. Generally, `mris_convert` is used to handle the conversion between different surface formats. For instance, in the directory `freesurfer/ernie/surf`, we can run

```
$ mris_convert ./lh.pial pial.stl
```

to create the file `lh.pial.stl` in the current directory.[3] The resulting file can be opened in several different programs, for instance, ParaView or Gmsh.

3.1.3 Creating a volume mesh from a surface

The third step of our initial meshing pipeline is to generate a mesh of the volume bounded by the surface representation. We will use the tailored package SVM-Tk (see Chapter 2.4.3) to convert from the STL surface representation to a volume mesh. The Python script below (`mri2fem/chp3/surface_to_mesh.py` in the book scripts) demonstrates the fundamentals of this process. The script (and all similar scripts in the following) can then be run from there as:

```
$ python surface_to_mesh.py
```

Recall that Python version 3 is required. If you have more than one version of python installed on your machine you may need to use 'python3' in the above, and all further commands, instead of the 'python' directive; this will explicitly specify which version of Python should be used to execute the script at hand. The script `surface_to_mesh.py` defines a Python function, named `create_volume_mesh`, within it that can be called as

```
create_volume_mesh("lh.pial.stl", "lh.mesh")
```

The function itself reads

```
import SVMTK as svmtk

def create_volume_mesh(stlfile, output, resolution=16):
```

---

[3] Note that it is possible to avoid the automatic addition of the prefix lh. in lh.pial.stl by adding the prefix ./ to the output filename.

```
# Load input file
surface = svmtk.Surface(stlfile)

# Generate the volume mesh
domain = svmtk.Domain(surface)
domain.create_mesh(resolution)

# Write the mesh to the output file
domain.save(output)
```

Given an input STL filename (`stlfile`), an output mesh filename with the `mesh` suffix (`meshfile`), and an optional mesh resolution, [4] the script creates an SVM-Tk `Domain` object, generates a volume mesh from the surface via the call to `create_mesh`, and saves this mesh in the `.mesh` format to the output mesh file. The mesh resolution parameter determines the maximum size of a tetrahedron in the volume mesh (relative to the overall bounding box length for the input surface): the higher the value, the higher the resolution – that is, the smaller the volume of each element in the volume mesh generated. Figure 3.2 shows meshes with `resolution = 16` (left) and `resolution = 64` (right). Mesh generation can be a costly operation, with higher run times (in the order of seconds to minutes) for higher resolutions.

The `.mesh` format is the standard mesh format for SVM-Tk and CGAL. However, it is not native to FEniCS, and so we need to convert the mesh to a FEniCS-supported file format (for example .xml, .xdmf, .h5). The Python package meshio (see Chapter 2.4.6) is well suited for this purpose and can convert between many different input and output mesh formats. For instance, to convert to the FEniCS .xdmf format, we can run

```
$ meshio-convert lh.mesh lh.xdmf
```

The .xdmf file can then be viewed in a program such as ParaView.

---

[4] The `resolution` parameter is a number that coarsely specifies the size of the mesh elements; higher values of `resolution` produce meshes composed of smaller average tetrahedral diameters. The `resolution` parameter default in this script is 16. Small changes in this value will not typically produce a visual difference (i.e. `resolution` =18) while larger differences (e.g. `resolution`=32 or 64) will produce visually distinct meshes with a clear difference in tetrahedral element size.

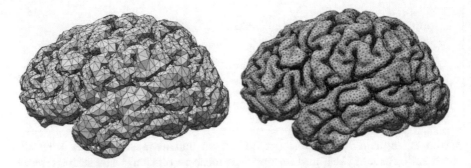

Fig. 3.2 Volume meshes of a left brain hemisphere produced by SVM-Tk from STL surface files, with lower (left) and higher (right) mesh resolutions.

3.2 Improved volume meshing by surface preprocessing

In the previous section, we stepped through the main pipeline for generating a volume mesh from MR images. However, the brain surfaces generated from T1 images often have a number of weaknesses:

- unphysiologically sharp corners,
- triangles with very large aspect ratios,
- topological defects such as holes, and
- a tendency to self-intersect or overlap with other surfaces.

These defects can cause the volume mesh generation to fail or result in low-quality meshes that are not suitable for numerical simulation. Therefore, surface preprocessing is often required to enhance surface quality prior to generating volumetric meshes. Here, we outline three main aspects of surface preprocessing: remeshing, smoothing and separation. The enhanced STL surfaces can then be directly inserted into the volume mesh generation as described in Chapter 3.1.3.

3.2.1 Remeshing a surface

To increase surface or volume mesh quality, we can remesh the original surface representation. The remeshing can, for instance, reduce the frequency of mesh cells that are overly distorted and reduce the density of vertices with large numbers of connected edges.

SVM-Tk includes utilities for remeshing surface files, and we can remesh our original surface `lh.pial.stl`, for example, via the following script (included as `mri2fem/chp3/remesh_surface.py` in the book scripts). We define a short Python function `remesh_surface` that can be called as

```
remesh_surface("lh.pial.stl", "lh.pial.remesh.stl", 1.0, 3)
```

The function itself reads

```
import SVMTK as svmtk

def remesh_surface(stl_input, output, L, n,
                   do_not_move_boundary_edges=False):

    # Load input STL file
    surface = svmtk.Surface(stl_input)

    # Remesh surface
    surface.isotropic_remeshing(L, n,
                                do_not_move_boundary_edges)

    # Save remeshed STL surface
    surface.save(output)
```

Here, we read the input STL file as an SVM-Tk `Surface`, and remesh using `isotropic_remeshing`, before saving the remeshed surface again as an STL file. We can specify more iterations and produce a finer mesh by increasing the integer value of n; we mention that n should not be thought of as an 'average inverse cell size' but only as a qualitative parameter such that higher values generally produce finer meshes. On the other hand, the floating point value of L corresponds to a quantitative mesh parameter: L indicates the maximum edge length of a mesh cell. Surfaces generated by FreeSurfer are typically in millimeters, and the volume meshes inherit this unit. The Boolean `do_not_move_boundary_edges` defines whether SVM-Tk is allowed to move the boundary vertices during the remeshing procedure (`False`) or not (`True`). It is generally advisable to allow boundary vertices to move, since requiring these to be fixed can cause the remeshing to fail.

Figure 3.3 shows the result `lh.pial.remesh.stl` with three remeshing iterations on the raw input file `lh.pial.stl`. The figure has been zoomed in to draw attention to local feature differences. Both files were viewed and visualized in ParaView.

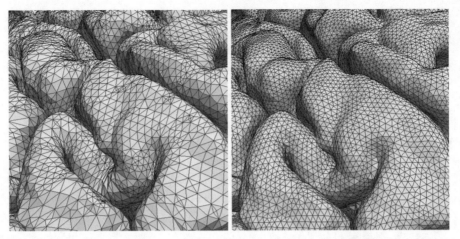

Fig. 3.3 Original (left) pial surface and (right) after remeshing with SVM-Tk. Remeshing improves the fidelity of the mesh and, at a high level, starts by subdividing tetrahedra based on a criterion, such as a maximal diameter or anisotropy condition, and further subdividing neighboring tetrahedron as needed (e.g. to ensure overall mesh quality).

3.2.2 Smoothing a surface file

To reduce the presence of non-physiological features such as sharp corners, it may be advantageous to smoothen the surfaces prior to volume meshing. SVM-Tk also includes utilities for smoothing surfaces, as we demonstrate in the script below (included as `mri2fem/chp3/smooth_surface.py` in the book scripts), using our remeshed surface `lh.pial.remesh.stl` as sample input. It is worth noting that surface smoothing operations alter the position of existing vertices; in particular, smoothing does not alter the number of elements in the mesh. Again, we define a short Python function `smoothen_surface` that can be called as

```
smoothen_surface("lh.pial.remesh.stl", "lh.pial.smooth.stl",
                 n=10, eps=1.0)
```

The function itself reads

```
import SVMTK as svmtk

def smoothen_surface(stl_input, output,
                     n=1, eps=1.0, preserve_volume=True):
```

```
# Load input STL file
surface = svmtk.Surface(stl_input)

# Smooth using Taubin smoothing
# if volume should be preserved,
# otherwise use Laplacian smoothing
if preserve_volume:
    surface.smooth_taubin(n)
else:
    surface.smooth_laplacian(eps, n)

# Save smoothened STL surface
surface.save(output)
```

The Boolean variable `preserve_volume` determines whether a shrinkage-preventing [64] Taubin smoothing (`True`) or a Laplacian smoothing (`False`) process should be used. From a conceptual point of view, Taubin smoothing is essentially a local smoothing iteration followed by a local 'swelling' operation that aims to prevent any shrinkage in the volume of the original patch, whereas Laplacian smoothing consists only of local smoothing operations and the volume of the original patch might not be preserved [64]. We recommend Taubin smoothing over the Laplacian approach since the latter tends to significantly diminish anatomical features. The integer value n sets the number of times the smoothing process should take place. Higher values will produce a smoother mesh; however, too a value that is too high may result in a loss of resolution in features on the brain surface, such as the sulci and gyri (grooves and bumps). Finally, the floating point value `eps` determines the strength of the smoothing operation for each smoothing iteration, and should be in the interval $[0, 1]$ with 0.0 (1.0) indicating no (full) smoothing.

Figure 3.4 shows the result of 10 iterations of smoothing; over-smoothing of the pial surface can lead to missing anatomical detail or errors (for example, Figure 3.4 (right)). It is generally advised to check the file visually by opening it directly, using either ParaView or Gmsh, to determine if more or less smoothing is needed; the number of iterations can vary between some surface STL files.

We can generate a higher quality mesh in XML or XDMF format by generating the volume mesh from the remeshed and smoothened STL surface using the Python call:

```
create_volume_mesh("lh.pial.smooth.stl", "ernie.mesh")
```

followed by:

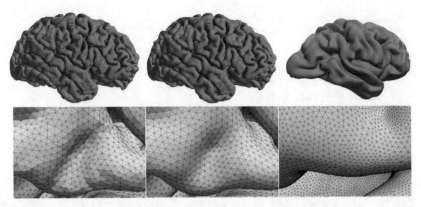

Fig. 3.4 Surface smoothing: original pial surface (`lh.pial.remesh.stl`, top left), after Taubin smoothing with SVM-Tk (`lh.pial.smooth.stl`, top middle), and Laplacian over-smoothing (top right). The bottom row shows a zoomed view of the three paradigms of the top row; the volume loss of the Laplace over-smoothing is evident.

```
$ meshio-convert ernie.mesh ernie.xml
$ meshio-convert ernie.xml ernie.xdmf
```

We will use the resulting `ernie.xdmf` in the simulations ahead in Chapter 3.3.

3.2.3 Preventing surface intersections and missing facets

SVM-Tk also includes utilities for repairing surface faults:

- Surfaces constructed by FreeSurfer can have topological defects, such as missing facets. Missing facets appear as 'holes' in the surface, i.e. a missing triangular simplex, when viewing a surface STL (.stl) file using ParaView or Gmsh. These defects can be repaired by following the FreeSurfer topological defects tutorial guide [3]. We can also attempt to fix missing surface facets using the SVM-Tk function `fill_holes`.
- The folds of pial surfaces can produce narrow gaps. Gaps that are shorter than the edges of the mesh may result in bridges instead of folds in the mesh, as exemplified in Figure 3.5. The function `separate_narrow_gaps` identifies narrow gaps and uses an algorithm to separate them based on the characteristics of the surrounding mesh. The function requires a negative value as input. Lower values of L results in a faster runtime for for the

algorithm, but may result in a more jagged surface. Higher values of L, e.g. those closer to zero, can extend computational time but generally result in a smoother surface.

- The command `collapse_edges` will combine short edges such that the new edge lengths are equal to the input target edge length.

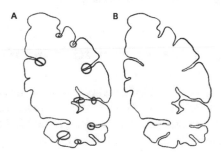

Fig. 3.5 Illustration of close junctures in a coronal slice of the pial surface created by FreeSurfer.

The following Python snippet illustrates the use of these commands (included as `mri2fem/chp3/svmtk_repair_utilities.py` in the book scripts):

```python
import SVMTK as svmtk

# Import the STL surface
lpial = svmtk.Surface("lh.pial.smooth.stl")

# Find and fill holes
lpial.fill_holes()

# Separate narrow gaps
# Default argument is -0.33.
lpial.separate_narrow_gaps(-0.25)
```

More commands involving multiple surfaces will be covered in Chapter 4.

3.3 Simulation of diffusion into the brain hemisphere

With a mesh representing the domain of interest, we are ready to start modeling and numerically simulating biophysical processes in this domain. As a first example, we will study diffusion into the brain parenchyma of a tracer injected in the subarachnoid space (SAS). This scenario might be encountered, for example, in clinical practice when gadobutrol is injected intrathecally (into the cerebrospinal fluid in the spinal cord SAS) [54], or in experimental research when fluorescent tracers such as dextran are injected in the cisterna magna of mice [33, 70]. Understanding the role of diffusion versus other mechanisms, such as convection, in the brain parenchyma is currently a hot topic in physiology and medicine [5].

3.3.1 Research question and model formulation

Let us ask the following question: assuming that solutes move in the brain parenchyma by diffusion alone, given an intrathecal injection of the contrast agent gadobutrol as in glymphatic MRI investigations [54], what evolution and distribution patterns of gadobutrol in the brain would we see up to 24 hours after injection?

To answer this, we define an initial boundary value problem for the diffusion of gadobutrol in the brain parenchyma. To do so, we need to prescribe the computational domain, initial conditions, parameter values, and boundary conditions for (1.1). It is also a good idea to think about the units when defining your simulation scenario. In brain mechanics at the tissue or organ level, millimeters (mm) or meters (m) and seconds (s) or minutes (min) are often appropriate units to use, although it is important to note that some physiological processes can take days, months, or years.

Here, we let the concentration u represent the (unitless) concentration, of gadobutrol tracer, solving (1.1) in Ω. The image-based mesh of the left brain hemisphere (for example `ernie.xdmf`) defines the computational domain Ω. Formally, Ω is then the union of the cells in the mesh. Note that the mesh defines the spatial coordinates of the domain and consequently the spatial unit. Meshes generated from FreeSurfer data are typically defined in terms of mm, and thus mm is the default unit for the spatial scale. The spatial unit can be redefined by rescaling the mesh, a simple operation in FEniCS. We also pick a final time and time unit, for instance $T = 1440$ min ($T = 24$ hours).

We assume that no gadobutrol is present in the domain initially, which translates to the initial condition:

$$u_0(x, 0) = 0 \quad \text{for all } x \in \Omega. \tag{3.1}$$

Next, we assume that gadobutrol is instantaneously distributed to the brain surface via the CSF in the SAS by setting the boundary condition:

$$u_d(x,t) = 2.813 \times 10^{-3} \quad x \in \partial\Omega, \quad t > 0 \tag{3.2}$$

where $\partial\Omega$ denotes the boundary of Ω, which represents the pial surface in this scenario. Clearly, the assumption that the concentration is known on the pial surface everywhere, and at all times, is overly simplistic. More realistic boundary conditions have been considered in the literature [17, 67], and one could also directly model the movement of tracer in the CSF in the subarachnoid space [30, 51]

Brain tissue is heterogeneous and anisotropic, which means that the effective diffusion coefficient D in (1.1) should vary in space (and probably in time over longer time scales) and be tensor valued. We will address these topics in Chapter 5, but for now we just consider a uniform and scalar-valued D. We estimate the average effective diffusion coefficient of gadobutrol in brain tissue to be:

$$D = 4.32 \times 10^{-1}\,\mathrm{mm}^2/\mathrm{hour} = 7.2 \times 10^{-3}\,\mathrm{mm}^2/\mathrm{min} \tag{3.3}$$

Finally, we assume that there are no sources or sinks of gadobutrol within the brain parenchyma, and thus set:

$$f(x,t) = 0 \quad x \in \Omega, \quad t > 0. \tag{3.4}$$

Again, this is clearly a simplification that does not account for potential exit pathways for gadobutrol from the parenchyma.

3.3.1.1 Quantities of interest

The computed solution u will vary in time and space and thus encodes a substantial amount of information. Initially, we are often interested in merely inspecting the solution visually or qualitatively. For more quantitative analysis and comparison with experimental or clinical findings, we typically compute quantities of interest associated with the solution. These quantities of interest can, for instance, be the total volume of solute in the entire domain over time:

$$Q(t) = \int_\Omega u(t)\,\mathrm{d}x, \tag{3.5}$$

the average concentration in local regions, or the concentration in specific points x over time: $u(x, t)$.

3.3.2 Numerical solution of the diffusion equation

To compute numerical solutions of the diffusion equation (1.1) in general, and our specific initial boundary value problem in particular, we will use a finite difference discretization in time and a finite element discretization in space [42, 24]. This is a common approach and we will implement this numerical scheme using FEniCS Project software [44, 9, 41]. As mentioned in our introductory remarks, we strongly encourage readers unfamiliar with numerically solving partial differential equations (PDEs) or FEniCS to study the FEniCS tutorial [41] before proceeding.

For the discretization in time, we define a set of discrete times $0 = t_0 \leq t_1 \cdots \leq t_N = T$, where N is the number of time steps and the time step (size) is $\tau_n = t_n - t_{n-1}$ for $n = 1, \ldots, N$. Our aim is to compute approximate solutions u_h^n of (1.1) such that $u_h^n \approx u(t_n)$ for each n. To this end, we introduce the (first-order, backward) finite difference approximation in time

$$u(t_n) \approx u^n, \quad u_t(t_n) \approx \frac{1}{\tau_n}(u^n - u^{n-1}), \qquad (3.6)$$

and obtain the time-discrete equations for $n = 1, \ldots, N$:

$$\frac{1}{\tau_n}(u^n - u^{n-1}) - \operatorname{div} D \nabla u^n = f(t_n) \quad \text{in } \Omega. \qquad (3.7)$$

Next, for the finite element discretization in space, we derive a discrete variational formulation of (3.7) by multiplying it by test functions ϕ belonging to a finite element space V defined relative to the mesh $\mathcal{T}$, integrating by parts, and moving all known terms to the right-hand side to obtain the following fully discrete problem: find the discrete solution $u_h^n \in V$ at each time $n = 1, \ldots, N$ such that

$$\langle u_h^n, \phi \rangle + \tau_n \langle \nabla u_h^n, \nabla \phi \rangle = \langle u_h^{n-1}, \phi \rangle + \tau_n \langle f^n, \phi \rangle, \qquad (3.8)$$

for all test functions $\phi \in V$, where we use the $L^2(\Omega)$-inner product notation

$$\langle u, \phi \rangle = \int_\Omega u \, \phi \, \mathrm{d}x. \qquad (3.9)$$

In addition, we require the discrete solution to satisfy the boundary condition (3.2), and to initially be given by the initial condition (3.1).

The fully discrete equation (3.8) is a good starting point for the FEniCS implementation of this scheme. We choose to approximate the solution using continuous piecewise linear finite element spaces.

3.3.3 Implementation using FEniCS

Our model problem is very similar to the heat equation problem presented in the FEniCS tutorial [41, Chapter 3.1], and we base our implementation on the algorithm and code presented there. We begin by importing the Python module `fenics`, and we also import `numpy` as a handy Python module for general numerics:

```
from fenics import *
import numpy
```

We then read the mesh that we have just generated into the FEniCS program. We use the XDMF mesh format and reader, since they are suitable for large-scale simulations and work seamlessly with MPI-parallel computing:

```
# Read the mesh
mesh = Mesh()
file = XDMFFile(MPI.comm_world, "ernie.xdmf") # mm
file.read(mesh)
file.close()
# Compute and print basic info about the mesh
print("#vertices =", mesh.num_vertices())
print("#cells =", mesh.num_cells())
print("max cell size (mm) =", mesh.hmax())
print("Volume (mm^3) = ", assemble(1*dx(mesh)))
```

We now define the parameters for the time discretization. We choose to simulate up to 72 hours and choose a time step τ (`tau`) of three min. Note that we use the Constant type to represent `time`; it is useful for updating functions depending on time later. In addition, it is good practice to keep track of the parameter units that are used. Here we just use the comments, although much more rigorous solutions, such as SymPy's unit systems [47], could be used.

```
# Define time discretization parameters
T = 72*60            # 72 hours in min
tau = Constant(3.0)  # Time step (min)
```

```
time = Constant(0.0)
```

We also define the diffusion coefficient, source function and initial condition (even though the latter two are just zero):

```
# Define the diffusion parameter
D = Constant(7.2e-3)   # mm^2/min

# Define the source function and initial condition
f = Constant(0.0)
u0 = Constant(0.0)
```

We now move on to consider the specification of the finite element discretization. We first define the finite element space V as the Lagrange elements of degree 1 defined relative to our mesh, and then define a `TrialFunction` and `TestFunction` over this space

```
# Define the finite element spaces and functions
V = FunctionSpace(mesh, "Lagrange", 1)
u = TrialFunction(V)
v = TestFunction(V)
```

We next define a `Function` to hold the value of the solution at the previous time step u_, and initialize this function with the initial condition u0:

```
# Define function to hold solution at previous time and
# assign initial condition to it
u_ = Function(V)    # AU (Arbitrary Unit)
u_.assign(u0)
```

Having defined these objects, we can express the variational problem (3.8), to be solved at each time step, in code. We also redefine u as a `Function` to represent the solution at the current time:

```
# Define the variational system to be solved at each time
a = (u*v + tau*D*dot(grad(u), grad(v)))*dx
L = (u_*v + tau*f*v)*dx

# (Re-)define u as the solution at the current time
u = Function(V)     # AU
```

Having defined the finite element space, we can also define the boundary condition, which will be imposed on the linear system of equations at each time step.

```
# Define the boundary condition: grow linearly up
# to the value c in the first 6 hours:
```

```
u_d = Expression("t < 6*60 ? t/(6*60)*c : c",
                  t=time, c=2.813e-3, degree=1)   # AU
bc = DirichletBC(V, u_d, 'on_boundary')
```

Note how we let the boundary value u_d depend on the previously defined
Constant time. This way, when time updates, so will u_d and bc.

We have now defined all the elements of the model problem and numerical
method and can start stepping through the solution algorithm. Since the bilin-
ear form on the left-hand side of (3.8) does not vary in time, we can assemble
this matrix once, outside the time loop, for the sake of efficiency:

```
# Assemble the left hand side matrix outside time loop
# for efficiency
A = assemble(a)
```

In order to view our solutions using ParaView, we define a PVD file (indicated
by the suffix .pvd) for storing the solution, computed at each time step, as
well as the initial data.

```
# Define file to store solutions and store initial solution
vtkfile = File("results/u.pvd")
u.assign(u_)
vtkfile << (u, float(time))
```

The PVD format works well for visualization in ParaView but is not a convenient format for reading the functions back into FEniCS; the XDMF or h5
formats are more appropriate for this.

Next, we compute the number of time steps, N, and initialize NumPy arrays
for storing computational quantities of interest at each time step, such as the
total amount of solute (amounts) cf. (3.5), and the concentration (concs) at
a specific point (p).

```
# Compute number of time steps and create arrays for
# computational quantities of interest
N = int(T/float(tau))
times = numpy.zeros(N+1)
amounts = numpy.zeros(N+1)
p = (-22.66, -48.23, 12.50)
concs = numpy.zeros(N+1)
```

Now we are ready to start stepping through time and to solve the finite
element system at each time step. We increment time with the timestep tau,
assemble the right-hand side into the vector b, apply the Dirichlet boundary
condition, solve the resulting linear system using an iterative solver (gmres,
amg), and update the previous solution with the current solution:

```
# Iterate over the time steps
for n in range(1, N+1):

    # Update time
    time.assign(time + tau)
    times[n] = float(time)

    # Assemble right-hand side
    b = assemble(L)

    # Apply boundary condition to linear system and solve it
    bc.apply(A, b)
    solve(A, u.vector(), b, "gmres", "amg")

    # Set previous solution to the current before moving on
    u_.assign(u)
```

Note that the use of "gmres", in the solve function above, tells FEniCS to use the generalized minimal residual method as the iterative solution technique; the use of "amg" specifies that the algebraic multigrid method should be used as the preconditioning approach for the iterative method. The interested reader can find more on the GMRES method, including convergence details, in [28] and a review of algebraic multigrid in [62]. Both GMRES and AMG are well suited to parabolic problems like (1.1), and FEniCS supports several alternative iterative methods and preconditioning options [41].

We can compute the total amount of solute by integrating the concentration over the domain, and the concentration in the specified point by evaluating the computed solution at this point. We also store the entire solution to the PVD file. Writing the solution to file can take some time, and, for fine time steps, it may often be practical to do this only, say, every second or tenth or one-hundredth time step:

```
    # Compute the total amount of solute and concentration
    # at given point:
    amounts[n] = assemble(u*dx)  # AU mm^3
    concs[n] = u(p)              # AU

    # Output progress and store solution every 10th time step:
    if n%10==0:
        print("Storing at n = %d (of %d), t = %g (min)"
              % (n, N, time))
        vtkfile << (u, float(time))
```

It can also be practical to store the computed quantities in a file, from which the contents may be read and plotted at a later time.

```
# Store amounts, times and concs to file
numpy.savetxt("results/times.csv", times, delimiter=",")
numpy.savetxt("results/amounts.csv", amounts, delimiter=",")
numpy.savetxt("results/concs.csv", concs, delimiter=",")
```

The Python module matplotlib, which offers quick and convenient plotting, can be used to plot the quantities of interest (the total amount of solute and the concentration at the given point) at different time points and save these in a format such as PNG. The resulting plots are shown in Figure 3.6. The complete script (including the plot code) is available in the book scripts (`mri2fem/chp3/diffusion.py`). The concentration at various points [5] can be investigated by modifying the value of p in the script, as shown in the code block below:

```
# Compute number of time steps and create arrays for
# computational quantities of interest
N = int(T/float(tau))
times = numpy.zeros(N+1)
amounts = numpy.zeros(N+1)
p = (-22.66, -48.23, 12.50)
concs = numpy.zeros(N+1)
```

The plot of total solute (Figure 3.6, left) shows a steady increase of solute within the brain as the available tracer at the boundary (implemented by the concentration boundary condition u_d) increases. As the process is diffusive, the total solute will increase until the concentration of the solution within the brain and at the boundary are in equilibrium. The approximation of the solute concentration (Figure 3.6, right) at a given point over time demonstrates an interesting numerical artifact: the concentration drops below zero to become negative between 0 and 550 min. Negative solute concentrations, which are clearly unphysiological, are a common numerical problem in diffusion simulations. A partial and often used remedy that can be applied within the context of diffusion simulations in the brain [17], is *mass lumping*. Mass lumping can reduce spurious negative concentrations, but may worsen the overall convergence of the numerical solutions. We will return to this issue in Chapter 6.

---

[5] The point we consider here has no particular significance; p was selected simply as an arbitrary point in the interior of the brain mesh.

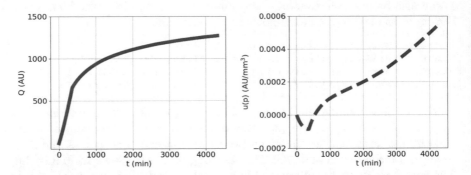

Fig. 3.6 Plots of the quantities of interest, in arbitrary units (AU), associated with the computed concentration: total amount of solute Q over time t (left) and concentration $u(p, t)$ for a given point p over time t (right). The unphysiological early concentration (right) results from a numerical artifact that is common when solving diffusion problems; in Chapter 6, we discuss a way to adjust our solution method to remedy this behavior.

3.3.4 Visualization of solution fields

To visualize computed solution fields, for instance, the concentrations stored in u.pvd (and the associated u*.vtu files) in the previous section, we will use ParaView. After launching the ParaView graphical user interface (see Chapter 2.4.5), we can open collections of files by $\boxed{\text{File} \to \text{Open}}$ and selecting the .pvd or .vtu file(s) from the folder results. ParaView is a powerful and versatile visualization tool, and we refer the reader to the extensive resources available on the ParaView website [4], including guides and tutorials. In Figure 3.7, we used ParaView to plot clips of snapshots of the solute concentrations, with the x-axis as the normal direction for the clips and the view direction, rescaled to the data range over all timesteps, and using the viridis color scheme.

3.4 Advanced topics for working with larger cohorts

We have now covered the entire computational pipeline from MR images to numerical simulation and visualization for a single imaging modality, a single stack of MRI data, and a single simulation scenario. In this section, we will

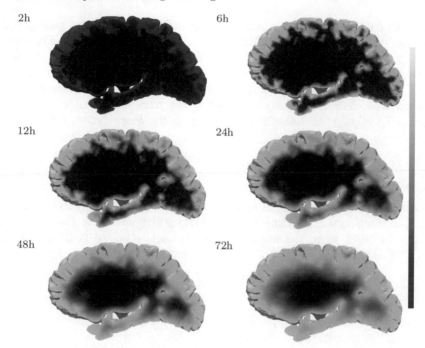

2h 6h

12h 24h

48h 72h

Fig. 3.7 Simulated tracer concentration at given times (2, 6, 12, 24, 48, and 72 hours) post-injection into subarachnoid CSF. Blue represents lower values, while yellow represents higher values. The scalar concentration is plotted on the whole brain volume mesh; the mesh has been sliced, in ParaView, for sagittal visualization.

cover some more advanced topics that are useful when it comes to working with more complex DICOM data collections.

3.4.1 Scripting the extraction of MRI series

When processing larger DICOM datasets consisting of many patients and/or studies, the extraction of single MRI series via a graphical interface can become tedious and error-prone. An alternative approach is to script the extraction of specific MRI series using FreeSurfer command line utilities.

The extraction process consists of two command line steps: probing the dataset for a given tag name and then extracting all data with the given tag

from the database. We will utilize the FreeSurfer command mri_probedicom
to examine and probe the DICOM metadata, so please ensure that FreeSurfer
is installed and configured (see Chapter 2.4.1) before proceeding. The useful
command mri_probedicom can be used to compare the metadata of different
DICOM files with the flag --compare followed by two DICOM filenames. We
can also view individual images associated to a particular DICOM file collection by specifying the image name and using the flag --view, for example:

```
$ cd dicom/ernie/DICOM
$ mri_probedicom --i IM_0162 --view
```

Alternatively, we may want to look for a specific tag. DICOM stores tags as
numeric identifiers. If we know that our MR scanner saves the *Protocol Name*
to tag 18 1030, as in the book DICOM dataset, then we could probe the
DICOM data for tag 18 1030 with the command:

```
$ mri_probedicom --i IM_0162 --t 18 1030
T13D
```

The complete description of possible options to mri_probedicom can be viewed
by using the --help flag.

To extract files with a specific tag on the command line, we can thus
probe each file, and copy files with a specific tag to a new directory. This
can be accomplished via the following bash script for example (also available
at mri2fem/chp3/extract_dicom.sh). The script takes three types of input,
namely, the DICOM directory, the output directory and an identifier for the
protocol name:

```bash
#!/bin/bash
# 1st argument $1: input DICOM folder
# 2nd argument $2: the output copy directory
# 3rd argument $3: the identifier

# Find all files in the directory and subdirectories
files=$(find $1 -type f )
for j in ${files}; do
    # Probe for protocol name (18 1030)
    name=$(mri_probedicom --i ${j} --t 18 1030)

    # Check if identifier is part of protocol name.
```

```
if [ "${name/$3}" != "$name" ]
    then
    # Copy file to (new) subdirectory
    mkdir -p ${2}/${name//[[:blank:]]/}
    cp ${j}  ${2}/${name//[[:blank:]]/}
    fi
done
```

This script uses the bash command **find** and the flag **-type f** to find all files in the input directory and its subdirectories. The script will go through all the files and probe each for the protocol name and check if the protocol name contains the identifier argument. Each file with the identifier in the protocol name will be copied to a folder named by the protocol name in the output directory. Spaces are removed from the protocol name, which is preferred to avoid errors when using FreeSurfer. Thus, we can use the script to extract all the images with the T1 string in the protocol name from the sample dataset:

```
$ cd dicom/ernie
$ ./extract_dicom.sh ./DICOM ./ "T1"
```

3.4.2 More about FreeSurfer's `recon-all`

The command **recon-all** is the primary command for FreeSurfer, since it will start the segmentation process. We have already described the necessary flags for this command, but below we continue the description, at an introductory level, with additional options. For more in-depth information, the interested reader can use the flag **-help**, which will print detail about the entire process and provide references to related articles and texts.

The command **recon-all** is a sequential process that consists of three steps comprising 34 different stages [3]. Data will be produced throughout the process, and are often required as input for the next stage. The recon-all command can be used to execute the full set of 34 stages or to execute only a portion of the stages. Some of the available **recon-all** flags are shown below; a full list of options is available online. [6]

- recon-all -autorecon1 -subjid *subject*: starts the step-1 process, which includes stages 1 to 5, involving normalization and skull stripping using the data in the subject folder named *subject*;
- recon-all -autorecon2 -subjid *subject*: starts the step-2 process, which includes stages 6 to 23, involving segmentation and surface generation using the data in the subject folder named *subject*;

---

[6] See https://surfer.nmr.mgh.harvard.edu/fswiki/recon-all/

- `recon-all -autorecon3 -subjid` *subject*: starts the step-3 process, which includes stages 24 to 34, involving statistical data generation and final parcellation using the data in the subject folder named *subject*;

These flags can be useful for restarting the segmentation. For instance, if a failure occurred at stage 34, then we can start over from stage 24 rather than from the beginning by using the flag `-autorecon3`, as shown above. This approach can also be useful when it is necessary to rerun the segmentation process after correcting an error. In FreeSurfer, there exist two types of errors: hard and soft errors. Hard errors will terminate the segmentation process, while soft errors are errors that we find in the produced data. Soft errors are mostly segmentation errors, such as the inclusion of the dura in the segmentation and erroneous segmentation of white matter. In this situation, we could edit the segmentation to correct the error, and run `recon-all` with the flag `-autorecon2-pial`. Specific details for the `-autorecon2-pial` flag, which we do not discuss here, can be found in the FreeSurfer documentation [3]. This will create new surfaces based on the corrected segmentation files. The correction of soft errors will not be covered further in this book. Instead, we refer to the FreeSurfer documentation [3].

We continue with the flag `-sd`, which can be used to specify the subject directory for the `recon-all` command. This can be quite useful for separating the segmentation data for different cohorts. The segmentation of CSF-filled structures, such as the ventricular system, may require the additional use of T2-weighted MR images to obtain an acceptable quality. We can include T2 MR images with the flag `-T2`, and we can use the flag `-T2-pial` to use the T2 MRI in the construction of pial surfaces.

The segmentation in FreeSurfer is based on the segmentation atlas of healthy subjects; therefore, the segmentation can often encounter hard errors for patients with abnormal brain anatomy. We can often allow the segmentation to finish if we add the flag `-notalairach`, which causes `recon-all` to skip assertion points in the first step. The log of `recon-all` is documented in the folder **scripts** and all the specific command lines can be found in **touch**.

Chapter 4
Introducing heterogeneities

In this chapter, we will consider how to mark, remove, and mesh different regions of the brain and its environment based on FreeSurfer segmentations. We will

- create hemisphere meshes differentiating between gray and white matter,
- create hemisphere meshes without ventricles,
- create brain meshes by combining the two hemispheres,
- map parcellations [1] onto brain meshes, and
- locally refine parcellated brain meshes.

4.1 Hemisphere meshing with gray and white matter

Gray and white matter differ substantially in their characteristics. These differences can often be represented in mathematical models and simulations by differing material properties. For instance, denoting the domains occupied by gray and white matter by Ω_g and Ω_w, respectively, we may want to consider heterogeneous diffusion tensors in (1.1), for example, such that

---

[1] A parcellation is a way of dividing the brain into distinct regions. FreeSurfer, for instance, does this as part of the `recon-all` process and this is why we can use Freeview to view the different parts of a subject's brain, such as the hippocampus or anterior cingulate, after `recon-all` completes.

$$D = D(x) = \begin{cases} D_g & x \in \Omega_g, \\ D_w & x \in \Omega_w \end{cases} \qquad (4.1)$$

where D_g and D_w take on different values, and D_g may be scalar-valued and D_w tensor-valued. To represent fields such as D in numerical simulations, it is useful to transfer the information about gray and white matter from the magnetic resonance (MR) images into the meshes. To introduce the basic concepts of differentiating brain regions, we will once more create a computational mesh of the left hemisphere. In Chapter 3, all of the mesh tetrahedra belonged to a single region. In this chapter, we extend the previous approach by creating a mesh where the individual tetrahedra will be labeled as belonging to the gray matter, the white matter, or the ventricles. In short, we will

- create STL files from the pial and white FreeSurfer left hemisphere surfaces,
- create a mesh from these STL files conforming to the interior interfaces between white and gray matter, using SVM-Tk, and
- include tags for different regions of the mesh. That is, for each tetrahedron in the mesh we want to label it as residing in the 'gray matter', 'white matter', 'ventricles', etc.

4.1.1 Converting pial and gray/white surface files to STL

Starting with the FreeSurfer segmentation and using the book data from `freesurfer/ernie/surf/` as an example, we first convert the left hemisphere pial surface (`lh.pial`) and gray-white interface surface (`lh.white`) files to the STL format (as described in Chapter 3.1.2):

```
$ mris_convert ./lh.pial pial.stl
$ mris_convert ./lh.white white.stl
```

We can now also improve the quality of the resulting surfaces as discussed in Chapter 3.2. [2]

---

[2] We leave the precise code for this as an exercise for the reader. The resulting files could be called `lh.pial.smooth.stl` and `lh.white.smooth.stl`, respectively. For simplicity, we assume that the resulting pial and white surface STL files are (re)named `lh.pial.stl` and `lh.white.stl`, respectively, in the following. As noted in Chapter 3, the automatic addition of the prefix lh., in lh.pial.stl, can be avoided by using ./ in the output filename.

4.1.2 Creating the gray and white matter mesh

Given these two surfaces, we can create a volume mesh conforming to the interior (gray–white) surface, with the white and gray regions identified separately. We will use SVM-Tk for this task, and proceed with an SVM-Tk code example. We will wrap the main functionality in a Python function `create_gw_mesh`. This function can then, for instance, be called as

```
create_gw_mesh("lh.pial.stl", "lh.white.stl", "ernie-gw.mesh")
```

to use the two STL surface files from above as input and create a new file `ernie-gw.mesh` for the resulting volume mesh.

Our function first loads the two surfaces using SVM-Tk:

```
import SVMTK as svmtk

def create_gw_mesh(pial_stl, white_stl, output):
    # Load the surfaces into SVM-Tk and combine in list
    pial  = svmtk.Surface(pial_stl)
    white = svmtk.Surface(white_stl)
    surfaces = [pial, white]
```

Notice that the list `surfaces = [pial, white]` contains two surfaces; the order of these surfaces in this list will matter. We next create a tailored SVM-Tk `SubdomainMap` object that represents a map between regions defined by surfaces and tags. This map is defined by (repeated) calls to `smap.add` with a string representing the region and an integer representing the tag as arguments. The (binary) string is a sequence of zeros and ones, with zero denoting the outside and one the inside. [3]

```
# Create a map for the subdomains with tags
# 1 for inside the first and outside the second ("10")
# 2 for inside the first and inside the second ("11")
smap = svmtk.SubdomainMap()
smap.add("10", 1)
smap.add("11", 2)
```

Intuitively, `smap.add("10",1)` will mean 'those tetrahedron inside surface 1 (pial surface) and outside surface 2 (white matter surface) should be marked

---

[3] Note that the STL surface format includes information about the orientation of the surfaces via the surface normals: each surface thus has an orientation, with an inside and an outside direction.

with the numeric (tag) value of 1' while `smap.add("11",2)` will mean 'those
tetrahedron inside surface 1 (pial surface) and inside surface 2 (white matter
surface) should be marked with the numeric (tag) value of 2'. At this point,
however, SVM-Tk does not know that the `surfaces` list and our `SubdomainMap`
object, `smap`, are related. Relating a surface list to a `SubdomainMap` object is
handled by creating a `Domain` object; a `Domain` object is constructed from the
ordered list of surfaces and the subdomain map as follows:

```
# Create a tagged domain from the list of surfaces
# and the map
domain = svmtk.Domain(surfaces, smap)
```

The `Domain` object reads the entries, registered above as `smap.add("10",1)`
etc, of the `SubdomainMap` with respect to the ordering of the entries in our
`surfaces` list. The important point here is that the order of the entries in
the `surfaces` list plays a key role; permuting the entries will yield different
results. For the code above, the string `"10"` is interpreted as 'inside pial' and
'outside white' and thus represents all tetrahedron in the region between the
pial and white surface; this region consists of only the gray matter. Similarly,
the string `"11"` is interpreted as 'inside pial' and 'inside white' and thus represents all tetrahedron in the white matter region since this region lies within
both surfaces. We will discuss `SubdomainMap` further in Chapter 4.1.3.

With `domain`, we can now create a volume mesh of suitable resolution and
save it in the .mesh format (as in Chapter 3.1.3):

```
# Create and save the volume mesh
resolution = 32
domain.create_mesh(resolution)
domain.save(output)
```

This script is also available as `mri2fem/chp4/two-domain-tagged.py` in the
book scripts, and can be run from there as:

```
$ python two-domain-tagged.py
```

As before, the resulting `.mesh` file can be converted to different formats using
meshio. For instance, to convert to the ParaView-friendly .vtu format, use:

```
$ meshio-convert ernie-gw.mesh ernie-gw.vtu
```

We used ParaView to visualize the tags associated with this mesh in
Figure 4.1. To see a view similar to that of Figure 4.1 in ParaView, first
load `ernie-gw.vtu` by selecting File→Open and s electing `ernie-gw.vtu`,

then click $\boxed{\text{Apply}}$ to show the mesh. In the left-hand window, under the
`Coloring` heading, select the second entry titled `medit:ref`. Now select
$\boxed{\text{Filters}\rightarrow\text{Alphabetical}\rightarrow\text{Clip}}$ from the top menu bar. The clipping plane
should, by default, appear in the middle of the mesh with the (red) clipping
plane in the sagittal orientation. Click the $\boxed{\text{Apply}}$ button in the left-hand
window to cut the mesh and reveal the tagged gray and white matter (tagged)
tetrahedron. We see that the mesh cells associated with the gray matter region
have the value 1, while the cells associated with the white matter region have
the value 2, as defined by our subdomain map.

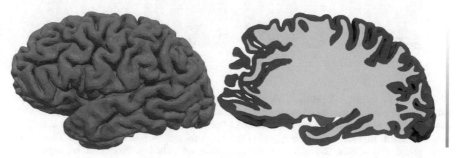

Fig. 4.1 Volume mesh of the left hemisphere conforming to the interior gray–white
interface. Gray matter is tagged with a value of 1 and white matter is tagged with
a value of 2. The color scale is inverted. A sagittal view (left) of the left hemisphere
volume mesh and a sliced (right) view exposes the interior white matter region.

As a side note, we observe that the ventricles have been labeled as white
matter (see Figure 4.1, right), since the ventricles are positioned inside the
gray–white interface surface. Removal of the ventricles from our computational
mesh is the topic of Chapter 4.2.

4.1.3 More about defining SVM-Tk subdomain maps

To delve into further detail about the SVM-Tk feature `SubdomainMap`, let's
consider another, more involved example. Let us now assume that we have
four surfaces: a left pial surface, a right pial surface, the whole white matter
surface and a surface enclosing the ventricles. A schematic of this set-up is

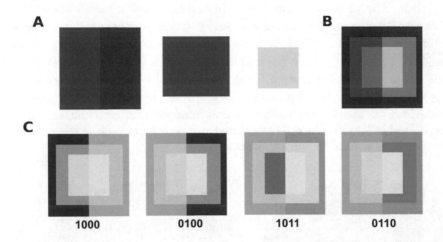

Fig. 4.2 The upper left panel, (A), shows three colored squares that differentiate the volumes enclosed by the large-scale surfaces files, in the `surfaces` list, at their simplest level. The left hemisphere volume is colored blue, the right hemisphere volume is colored red, the white matter volume is purple and the ventricle volume is yellow. The upper right panel (B) shows the complex combination of volumes that can be separately tagged using `smap.add` and the surfaces defined in the `surfaces` list. The regions of (B) are colored by an independent color scheme that shows all possible combinations of volumes addressable by `smap.add`. The bottom panel (C) shows four of the six subdomains of (B) and their corresponding bit strings. The left image, corresponding to bit string '1000', denotes the volume that is within the left hemisphere, but not within the right hemisphere, white matter or ventricles - hence the gray matter of the left hemisphere. The image with bit string '0100' is completely analogous for the right hemisphere. The bit string '1011' refers to the surface that is within the left hemisphere and within both the white surface and ventricles; hence the left ventricle. Finally, the image for the bit string '0110' corresponds to the white matter of the right hemisphere. For completeness, the remaining two bit strings for the regions shown in (B) are '1010' (left, magenta region) and '0111' (right, golden region).

illustrated in Figure 4.2. Now, let us see how we can tag all the tetrahedra of the ventricles.

To represent the subdomains, assuming that `lpial`, `rpial`, `white`, and `ventricles` exist as `svmtk.Surfaces`, we could use the following sample code:

```
surfaces = [lpial, rpial, white, ventricles]
smap = svmtk.SubdomainMap()
```

The ventricles are filled with cerebrospinal fluid. A practical reason to tag the ventricles separately may be, for example, to specify a much faster (isotropic) diffusion coefficient in the ventricular domain. Here, we demonstrate how to mark all of the mesh tetrahedra that lie inside the subdomain defined by the ventricle surface with a (tag) value of 6. To do this, we need to identify the placement of volume corresponding to the left ventricles; as we have seen, this volume is defined implicitly by the surfaces within the list `surfaces`. The ventricles in the left hemisphere are inside the left pial surface, outside the right pial surface, inside the gray-white matter surface and inside the ventricular surface, resulting in the bit-string '1011'. We could thus call `smap.add` as

```
smap.add("1011", 6)
```

Similarly, we can tag the ventricle volume within the right pial surface by

```
smap.add("0111", 6)
```

Finally, we can tag the ventricular volume at the intersection of the left and right pial surfaces via

```
smap:add("1111", 6)
```

Indeed, tagging the entire ventricular volume in this manner requires that we add all three of the lines above.

Finally, we note that SVM-Tk also allows for marking several domains at once:

```
smap.add("10*", 6)
```

will mark all underlying domains - that is, in this case '1000', '1001', '1010', and '1011'. Note that the use of the asterisk requires a prior specification of the number of input surfaces, either as an optional argument in the constructor, of a `SubdomainMap` object, such as `smap`, or by using the member function `set_number_of_surfaces` of the `SubdomainMap` class (for example, `smap.set_number_of_surfaces`). Therefore, as an alternative, one could just include the line

```
smap.add("*1", 6)
```

instead of adding each ventricular volume separately.

4.2 Separating the ventricles from the gray and white matter

The volume hemispheric meshes created in Chapter 3, and the volume hemisphere mesh illustrated in Figure 4.1 include the ventricles as part of the white matter. Since the physics of the fluid-filled ventricles and the soft but solid brain cerebrum may be very different, removal of the ventricles from the hemisphere volume is a useful operation. Here, we demonstrate how to (i) use FreeSurfer to extract and postprocess the ventricular surface, and (ii) remove the resulting ventricular volume from the cerebrum.

4.2.1 Extracting a ventricular surface from MRI data

We will extract the ventricle surface(s) from our T1 MRI data via FreeSurfer. Extracting a ventricular surface representation is relatively straightforward, while extracting a high-quality surface representation may be more involved. Accordingly, we will be introducing tools and utilities of increasing complexity.

Segmentations and parcellations: A sneak peek

Recall that FreeSurfer's `recon-all` generates a number of surface and volume files (see, for example, Chapter 3.1.2). In particular, the FreeSurfer-generated `mri/` directory includes volume-based data, such as T1-weighted images, segmentations, and parcellations. These volume files have the extension `mgz`, and the segmentations and parcellations can be identified by the base filename. For instance, the file `aseg.mgz` stands for automatic segmentation, and the file `wmparc.mgz` stands for white matter parcellation. The parcellation will split the segmentation into finer regions, for example, the cortical gray matter will be divided into 35 regions[4] for each hemisphere. We can use the seg-

---

[4] FreeSurfer defines regions via an anatomical atlas. An atlas is a labeling of distinct regions. The 35 regions referenced here correspond to the Desikan-Killiany atlas which ships with FreeSurfer; the Destrieux atlas also ships with FreeSurfer and can be used to annotate various cortical regions. More information regarding FreeSurfer atlas annotations is available at https://surfer.nmr.mgh.harvard.edu/fswiki/CorticalParcellation

mentation or the parcellation files to construct the surface of the ventricular volume.

The segmentations can be inspected using a tool such as Freeview. As an example, we use the FreeSurfer generated files from our dataset at freesurfer/ernie/mri, and visualize the aseg.mgz file:

```
$ cd freesurfer/ernie/mri
$ freeview --colormap lut --v aseg.mgz
```

The list in the left-hand panel of the Freeview window shows the segmentation tags, that is, the values associated with different brain regions. Alternatively, hovering the pointer over a voxel will cause the corresponding region tag and name to appear in the bottom right corner. For instance, the left hippocampus has tag 17, while the fourth ventricle has tag 15. We will look further into segmentations and parcellations in Chapter 4.4, including a visualization in Figure 4.6.

Extracting and binarizing voxel-based information

The FreeSurfer command mri_binarize is used to extract and mark voxels that contain a certain type of information such as a range of signal values or a collection of segmentation tags. The command includes about 40 optional flags, all of which are described in

```
$ mri_binarize --help
```

or via the FreeSurfer online documentation [3]; we will focus on but a few of these here.

The input file is given following the flag --i, and a volume output file (.mgz) is given following the flag --o. A surface output file (.stl) can be given in addition to or instead of the volume output following the flag --surf. The essential flag --match, followed by one or more integers, will mark all voxels whose assigned segmentation region identification tag matches any of the given integer values. For instance, to select all voxels from the fourth ventricle, we can use `--match 15` [5]. Alternatively, specific regions can also be identified by designated flags, for instance:

- `--ventricles` marks voxels in the third and lateral ventricles and in the choroid plexus,
- `--ctx-wm` marks voxels in the cerebral white matter,
- `--gm` marks voxels in the gray matter, and
- `--subcort-gm` marks voxels in the subcortical gray matter, including the gray matter in the cerebellum and brainstem.

These flags can be combined. For example,

```
$ mri_binarize --i aseg.mgz --ventricles --match 15 --o v.mgz
```

will mark the third and lateral ventricles (via the `ventricles` flag) and the fourth ventricle (with match value 15). In the output file, all the marked voxels will have the value one and the rest will be set to zero. This result can be changed by specifying the output bin value with the optional flag `--bin` followed by an integer. The marked voxels will now have the selected bin value in the output. The surface output flag `--surf` is often used together with the flag `--surf-smooth` followed by an integer determining the number of smoothing iterations on the output surface.

To extract the ventricular surface from our white matter parcellation, we define a customizable bash script (`extract-ventricles.sh`). We begin by defining the input and output file names:

```
#!/bin/bash

# Input and output filenames
input="wmparc.mgz"
output="ventricles.stl"
```

To extract a surface file of the ventricular system, we call `mri_binarize` with the input filename (in `input`), the `--ventricles` flag, additional tags given by `matchval`, and a number of smoothing iterations:

```
mri_binarize --i $input --ventricles \
```

---

[5] FreeSurfer assigns a numeric label to each identified region in the brain. These labels can be viewed by opening a subject's `aseg.mgz` file using Freeview. Doing so, we see that a value of '15' is assigned by FreeSurfer to the region that its segmentation procedure identifies as the subject's fourth ventricle.

```
        --match $matchval \
        --surf-smooth $num_smoothing \
        --surf $output
```

Prior to this call, we allow for the inclusion of the fourth ventricle and aqueduct by setting the `matchval` variable [6] and we set the `num_smoothing` iterations:

```
# Also match the 4th ventricle and aqueduct?
include_fourth_and_aqueduct=true
if [ "$include_fourth_and_aqueduct" == true ]; then
    matchval="15"
else
    matchval="1"
fi
num_smoothing=3
```

We suggest setting `num_smoothing` to an integer value between one and five. The resulting ventricular surfaces with zero and five smoothing iterations, both including the fourth ventricle, are shown in Figure 4.3.

In practice, the decision to include or discard the fourth ventricle and aqueduct is data specific. The aqueduct may not be well resolved in the MRI data on a patient-by-patient basis; if the aqueduct is not visible in the data then keeping the fourth ventricle leads to a ventricle system that is not connected, as Figure 4.3 indeed shows. Moreover, if we include the fourth ventricle and aqueduct, we should be cautious regarding the extent of the smoothing.

Improving the morphology of the ventricular surface

In Section 4.2.2, we will use the ventricular surface to modify a volume mesh. In this section, we discuss improving the ventricle surface by fixing a few morphological defects that may be present following the extraction process. In particular, we will introduce the FreeSurfer utilities `mri_volcluster` and `mri_morphology`. These tools can be used to: remove the smaller disconnected

---

[6] Setting the `matchval` variable only *allows for* the inclusion of the fourth ventricle and aqueduct. In particular, the aqueduct is a small, fine structure and is typically not fully differentiated. Note that volume files can be edited manually using Freeview, for instance to repair a partially resolved or missing aqueduct. See https://surfer.nmr.mgh.harvard.edu/fswiki/FreeviewGuide/FreeviewTools/VoxelEdit for more detail.

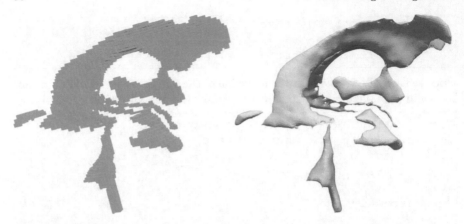

Fig. 4.3 Ventricular surfaces, including the fourth ventricle, extracted and generated by FreeSurfer from MRI images. No smoothing of the output surface (left) and five smoothing iterations (right). Note the disconnected regions.

ventricle regions that may appear in the original surface extraction; close small holes in the large ventricle surface; and smooth the resulting surface before further use.

- `mri_volcluster` is used to identify clusters in a volume. A cluster is defined as a set of continuous voxels that satisfies a specified volume threshold criteria; we will specify a minimum volume threshold in the code example below. The input file is given following the flag `--in`, `--thmin` gives a minimum threshold value, `--minsize` gives a minimal cluster volume (in mm³). Different output flags are admissible [3], including `--ocn`, used to save the output volume file with sorted clusters, where all voxels of the largest cluster will have the tag 1, the second largest would have the tag 2, and so on.
- `mri_morphology` is used to perform certain operations on volume files and supports many operations. These operations include opening, closing, dilating, eroding and filling holes in a volume. We will make use of the 'close' operation, of this utility, to close holes in an extracted ventricle domain.

Using these functions, we may extract an improved, higher-quality ventricular surface by removing apparently disconnected regions. Our algorithm takes the following steps:

1. We extract the lateral and third ventricles into a separate volume file.

2. In this separate volume, we extract clusters of connected voxels, but only those above a minimal cluster size. Since the average adult volume of cerebrospinal fluid in the ventricles is about 150 mm^3, we set the threshold size to be around 100 mm^3.
3. We extract the largest of these clusters (thus ignoring smaller, disconnected regions)
4. We close any holes in the resulting volume as necessary.
5. We extract the surface of the resulting volume and smooth it as necessary.

The corresponding continuation of our Bash script is as follows. Note how we output the clusters sorted by size using the argument --ocn to mri_volcluster and extract the largest cluster by matching on one in the subsequent call to mri_binarize. We allow for setting the number of closing iterations num_closing and the minimal largest cluster size V_min as parameters. We advise setting the number of closing iterations relatively low (e.g. to 1 or 2). Setting the number of closing iterations too high can cause non-physiological connections in the resulting ventricular surface.

```
# Other parameters
postprocess=true
num_closing=2
V_min=100

if [ "$postprocess" == true ]; then
    mri_binarize --i $input --ventricles \
                 --o "tmp.mgz"

    mri_volcluster --in "tmp.mgz" \
                   --thmin 1 \
                   --minsize $V_min \
                   --ocn "tmp-ocn.mgz"

    mri_binarize --i "tmp-ocn.mgz" \
                 --match 1 \
                 --o "tmp.mgz"

    mri_morphology "tmp.mgz" \
                   close $num_closing "tmp.mgz"

    mri_binarize --i "tmp.mgz" \
                 --match 1 \
```

```
                      --surf-smooth $num_smoothing \
                      --surf $output

     rm tmp.mgz
     rm tmp-ocn.mgz
     exit
fi
```

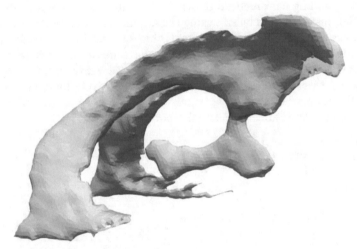

Fig.   4.4 Post-processed   ventricle   surface   extracted   from   MRI   using
FreeSurfer. This surface file (`ernie-ventricles.stl`) is created by the script
`mri2fem/chp4/extract-ventricles.sh`.

Figure 4.4 shows the ventricular surface STL file generated by the above code,
with `postprocess=true`, visualized in ParaView.

4.2.2 Removing the ventricular volume

In this section, we demonstrate how to remove a subvolume defined by an enclosing surface. Though we will focus here on removing the volume enclosed by
the ventricle surface, as extracted in the previous section, the general process
will also work for any volume defined by a closed surface STL file. The core

idea is to use SVM-Tk to generate tags for the different subvolumes in the domain and then simply delete the volume corresponding to a specific tag.

We assume that we have the left pial, gray/white matter, and ventricular surfaces available as STL files. Again, we will wrap the main functionality in a Python function, called create_gwv_mesh, to create the 'gray matter plus white matter' volume mesh. This function can then, for instance, be called as

```
create_gwv_mesh("lh.pial.stl", "lh.white.stl",
                "lh.ventricles.stl",
                "lh.no-ventricles.mesh")
```

This code example is included in mri2fem/chp4/three-domain-tagged.py.

We first create Surfaces from the surface STL files:

```
import SVMTK as svmtk

def create_gwv_mesh(pial_stl, white_stl, ventricles_stl,
                    output, remove_ventricles=True):

    # Create SVMTk Surfaces from STL files
    pial  = svmtk.Surface(pial_stl)
    white = svmtk.Surface(white_stl)
    ventricles = svmtk.Surface(ventricles_stl)
    surfaces = [pial, white, ventricles]
```

We then tag different regions using SubdomainMap:

```
    # Define identifying tags for the different regions
    tags = {"pial": 1, "white": 2, "ventricle": 3}

    # Define the corresponding subdomain map
    smap = svmtk.SubdomainMap()
    smap.add("100", tags["pial"])
    smap.add("110", tags["white"])
    smap.add("111", tags["ventricle"])
```

As before, we create a tagged mesh (of a given resolution) of the domain via the surfaces and the subdomain map:

```
    # Mesh and tag the domain from the surfaces and map
    domain = svmtk.Domain(surfaces, smap)
    resolution = 32
    domain.create_mesh(resolution)
```

We can now, via a call to SVM-Tk remove_subdomain, remove the mesh cells tagged as within the ventricles before saving the mesh:

```
# Remove subdomain with right tag from the domain
if remove_ventricles:
    domain.remove_subdomain(tags["ventricle"])

# Save the mesh
domain.save(output)
```

Note that `remove_subdomain` can also handle the removal of multiple subdomains by providing a tuple of tags as input. The resulting meshes, with and without ventricles removed, are shown in Figure 4.5.

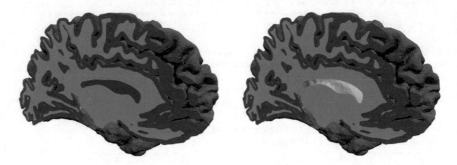

Fig. 4.5 Volume meshes of the left hemisphere (sagittal planes), conforming to the gray matter, white matter, and ventricles, with ventricles marked in blue (left) and with ventricles removed (right). Note that the tetrahedral boundary lines of the mesh have been suppressed for visual clarity. To view the tetrahedra of the mesh, select the `Surface With Edges` option, for the `Representation` setting in the left-hand pane, after loading the `.mesh` files, created by `three-domain-tagged.py`, in ParaView.

4.3 Combining the hemispheres

In this section, our aim is to create a mesh that includes both the left and right hemispheres, with gray and white regions tagged, and the ventricular volume removed. We will combine the approaches of the previous sections with SVM-Tk techniques for working with the union of multiple surfaces.

4.3.1 Repairing overlapping surfaces

FreeSurfer generates the right and left hemisphere surfaces separately. Combining surfaces from different hemispheres can therefore create problems, such as:

- The hemisphere surfaces overlap, creating bridges in the cortical gray matter, at the mesh and surface level, that do not exist physically.
- The hemisphere surfaces may have gaps between them which are too large and are unphysical. In this case, the resulting mesh may have undesirable gaps between the hemispheres where they would otherwise be connected by the white matter nerve tracts.

In general, we want to join the hemisphere surfaces, via the white matter nerve tracts, while simultaneously avoiding overlapping surfaces in the cortical gray matter. SVM-Tk includes utilities to address such challenges. In particular:

- `separate_overlapping_surfaces` can separate overlapping surfaces,
- `separate_close_surfaces` can separate nearly overlapping surfaces, and
- if we desire a single surface for the white matter but the white matter surfaces only partially overlap, due to smoothing for instance, the SVM-Tk function `union_partially_overlapping_surfaces` offers an improved set of features to handle the union operation of the white matter surfaces.

The following code snippet shows how these functions are used:

```
# Input Surfaces
rpial = svmtk.Surface("rh.pial.stl")
lpial = svmtk.Surface("lh.pial.stl")
rwhite = svmtk.Surface("rh.white.stl")
lwhite = svmtk.Surface("lh.white.stl")

# Create white matter surface as union of hemispheres
white = svmtk.union_partially_overlapping_surfaces(rwhite,
                                                   lwhite)

# Separate overlapping and close vertices between
# the left and right pial surfaces,
# but only outside the optional third argument, which
# in this example is the white surface:
svmtk.separate_overlapping_surfaces(rpial, lpial, white)
svmtk.separate_close_surfaces(rpial, lpial, white)
```

4.3.2 Combining surfaces to create a brain mesh

We assume that left pial, left white, right pial, right white and ventricular surface STL files have been extracted, converted and possibly processed, by the processes described in the previous section. Now, how do we combine these to create a complete brain mesh? Again, we proceed via an SVM-Tk code example (with the complete code: `mri2fem/chp4/fullbrain-five-domain.py`). We wrap the main functionality in a Python function `create_brain_mesh`. This function can then, for instance, be called as:

```
stls = ("lh.pial.stl", "rh.pial.stl",
        "lh.white.stl", "rh.white.stl",
        "lh.ventricles.stl")
create_brain_mesh(stls, "ernie-brain-32.mesh")
```

We begin by loading Surfaces from the STL files.

```
import SVMTK as svmtk

def create_brain_mesh(stls, output,
                      resolution=32, remove_ventricles=True):

    # Load each of the Surfaces
    surfaces = [svmtk.Surface(stl) for stl in stls]
```

We take the union of the left and right white surfaces to illustrate the possibility of combining surfaces:

```
    # Take the union of the left (#3) and right (#4)
    # white surface and put the result into
    # the (former left) white surface
    surfaces[2].union(surfaces[3])

    # ... and drop the right white surface from the list
    surfaces.pop(3)
```

It is natural to ask whether we can do the same with the pial matter; indeed, the union of the left and right pial surfaces is possible. However, whether the joined surface can be successfully meshed without further postprocessing is data specific. There is a higher chance that the FreeSurfer segmentation process of the left and right pial MRI surface data can lead to non-physical intersections, thus producing left and right pial surface files that overlap and self-intersect when combined. Self-intersections within a surface can then cause the meshing process to fail. Similarly, one could simply work with all five surfaces separately. This approach leads to a more complex tagging process, with `SubdomainMap`, but alleviates difficulties, such as the self-intersections mentioned above, that can arise when computing the union of two surface objects.

We continue by creating tags, a subdomain map, domain and mesh, and leave the option of removing the ventricles, as in previous examples:

```python
# Define identifying tags for the different regions
tags = {"pial": 1, "white": 2, "ventricle": 3}

# Label the different regions
smap = svmtk.SubdomainMap()
smap.add("1000", tags["pial"])
smap.add("0100", tags["pial"])
smap.add("1010", tags["white"])
smap.add("0110", tags["white"])
smap.add("1110", tags["white"])
smap.add("1011", tags["ventricle"])
smap.add("0111", tags["ventricle"])
smap.add("1111", tags["ventricle"])

# Generate mesh at given resolution
domain = svmtk.Domain(surfaces, smap)
domain.create_mesh(resolution)

# Remove ventricles perhaps
if remove_ventricles:
    domain.remove_subdomain(tags["ventricle"])

# Save mesh
domain.save(output)
```

Note that running this code example as is will take a few minutes. The resulting mesh can be visualized using ParaView after conversion from .mesh to .vtu or .xdmf.

4.4 Working with parcellations and finite element meshes

FreeSurfer's `recon-all` automatic segmentation process can identify almost two hundred different brain regions.[7] FreeSurfer labels identified regions with a numeric code. For instance, in the left hemisphere, FreeSurfer assigns the numeric code of 17 to the hippocampus[8], 1035 to the gray matter insula, 3035 to the white matter insula, 1028 to the gray matter superiorfrontal region and 3028 to the white matter superiorfrontal region. Figure 4.6 (left) illustrates some of these regions using `freesurfer/ernie/mri/wmparz.mgz` as an example.

4.4.1 Mapping a parcellation onto a finite element mesh

In a brain parcellation, each region is identified by an integer value. Our current goal is to map these region tags onto the generated volume mesh and into a FEniCS-compatible format. Doing so involves:

- reading and working with image (voxel-based) data in Python,
- representing discrete mesh data in FEniCS,
- mapping values from voxel indices/voxel space to mesh coordinates - that is, left-to-right, posterior-to-anterior, inferior-to-superior (RAS) space.

As usual, we will illustrate these steps using a concrete code example (included in `mri2fem/chp4/map_parcellation.py`). We wrap our main functionality in

---

[7] FreeSurfer's `recon-all` parcellates a brain into the regions defined by the Desikan-Killiany and Destrieux atlases. Numbers are assigned to the different regions, by default, when FreeSurfer performs the segmentation. These numbers are specific to FreeSurfer but independent of a subject. We note that the particular brain matter that FreeSurfer identifies as 'region N' (where N is some number) may differ between patients depending on their individual brain topology.

[8] In many neuroscience applications, the hippocampus region is of special interest because it is central to memory consolidation. The reader is refered to the interesting story of Henry Molaison, aka Patient H.M., [61, 57] who lost his ability to form new memories after the removal of both the left and right hippocampus. The procedure was performed as a treatment for his epilepsy. He lived for more than 50 years after the removal of his hippocampus and participated voluntarily in many scientific experiments, demonstrating the crucial role of the hippocampus. He never recognized the scientists who frequently visited him.

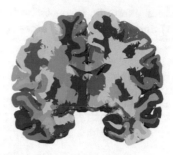

Fig. 4.6 Brain parcellations: (left) as generated by FreeSurfer and visualized using Freeview and (right) the same parcellation transferred onto the FEniCS brain mesh and visualized using ParaView (with different colors, different slices, and different view angles).

a Python function `map_parcellation_to_mesh` taking the parcellation filename and mesh filename as input (with `wmparc.mgz` and `ernie-brain-32.xdmf` from the previous section as an example).

```
map_parcellation_to_mesh("wmparc.mgz", "ernie-brain-32.xdmf")
```

Working with image data in Python

We will use the Python packages NiBabel to work with neuroimaging data, NumPy for general numerics in Python, and FEniCS to represent the mesh and the mesh data:

```
import numpy
import nibabel
from nibabel.affines import apply_affine
from dolfin import *
```

We begin by loading the image data from the parcellation file as follows:

```
def map_parcellation_to_mesh(parcfile, meshfile):
    # Load image from the parcellation file,
    # and extract the data it contains
    image = nibabel.load(parcfile)
    data = image.get_fdata()
```

We can, for example, inspect the image shape (number of voxels in each dimension) and extract voxel values by indexing the `data` array:

```
# Examine the dimensions of the image and
# examine the tag for the voxel located at
# data position 100, 100, 100
print(data.shape)
print(data[100, 100, 100])
```

Representing discrete mesh data in FEniCS

We aim to map the parcellation labels, generated by FreeSurfer during segmentation, onto the brain mesh. At this point, we have loaded the FreeSurfer-generated `wmparc.gz`; this parcellation was created from a set of T1 MR images. We will also work with a mesh file: `ernie-brain-32.xdmf`. This mesh file was generated from surfaces extracted from files that were also constructed by FreeSurfer from a set of T1 images. We mention this because the following fact is important: to label a mesh file with parcellated region IDs, it is imperative that the T1 images used to generate the parcellation and the T1 images used to generate the mesh file are the same set of images.

One way to associate discrete data, such as a parcellation label, with mesh elements is to use a FEniCS `MeshFunction`. Mesh functions can be associated with geometrical objects, X, of various dimensions, $d = \dim(X)$. We can associate mesh functions with cells ($d = 3$), faces ($d = 2$), edges ($d = 1$), or vertices ($d = 0$). Here, we aim to identify the parcellation region for each cell in the brain mesh and will thus make use of a cell function. We first import the full brain mesh:

```
# Import brain mesh
mesh = Mesh()
with XDMFFile(meshfile) as file:
    file.read(mesh)
print(mesh.num_cells())
```

Next, we create a mesh function for the mesh entities of dimension 3 (tetrahedral mesh cells). Our `MeshFunction` object will define a function whose input is a tetrahedron of the mesh and whose output, or associated value, is a real number; we start by specifying a mesh function associating an initial value of zero to all tetrahedra of the mesh:

```
# Define mesh-based region representation
n = mesh.topology().dim()
```

```
regions = MeshFunction("size_t", mesh, n, 0)
print(regions[0])
print(regions.array())
```

Note that the `MeshFunction` can be indexed directly, or its values can be accessed via the member function `array`.

Our strategy now is: iterate over all cells in the mesh; identify the parcellation region label for each mesh cell; and set the associated value of the mesh function for that cell to the corresponding parcellation label value. However, at this point, a key question arises: While we can index the image data by its indices and the mesh by its cell or vertex indices (and/or their coordinates), how can we identify the voxel index corresponding to a given cell or vertex (coordinate), and vice versa?

Converting between indices, coordinates, and spaces

Converting between voxel indices and other coordinates is a core problem that we will encounter several times. To address this task, we begin by introducing some nomenclature. The span of the image dimensions is often referred to as the (T1) *voxel space*, with indices (or dimensions) (i, j, k). On the other hand, the span of the mesh axes defines the RAS (left-to-right, posterior-to-anterior, inferior-to-superior) space with coordinates (x, y, z).

We aim to construct the transformation f from voxel space to RAS space:

$$(x, y, z) = f(i, j, k), \qquad (4.2)$$

as well as its inverse, f^{-1}, from RAS to voxel space. Fortunately, the parcellation information allows us to extract this transformation easily:

```
# Find the transformation f from T1 voxel space
# to RAS space and take its inverse to get the
# map from RAS to voxel space
vox2ras = image.header.get_vox2ras_tkr()
ras2vox = numpy.linalg.inv(vox2ras)
```

Note that FreeSurfer operates with several coordinate systems, including different RAS spaces. In this book, RAS space will refer to the FreeSurfer surface RAS space, which is identified via the suffix `tkr` in the code snippet above.

Next, we iterate over the cells of the mesh. For each cell, we will: extract the cell index; extract the RAS coordinates of the cell midpoint; convert from the RAS coordinates to voxel space (via a call to `apply_affine`); round off to the nearest integer values to find the image indices; and map the corresponding image data values into the **regions** mesh function. The code for this iterative process is

```
print("Iterating over all cells...")
for cell in cells(mesh):
    c = cell.index()

    # Extract RAS coordinates of cell midpoint
    xyz = cell.midpoint()[:]

    # Convert to voxel space
    ijk = apply_affine(ras2vox, xyz)

    # Round off to nearest integers to find voxel indices
    i, j, k = numpy.rint(ijk).astype("int")

    # Insert image data into the mesh function:
    regions.array()[c] = int(data[i, j, k])
```

We save the resulting mesh function in XDMF format (suitable for ParaView) and in the HDF5 format (suitable for further FEniCS processing):

```
# Store regions in XDMF
xdmf = XDMFFile(mesh.mpi_comm(),
                "results/ernie-parcellation.xdmf")
xdmf.write(regions)
xdmf.close()

# and/or store regions in HDF5 format
hdf5 = HDF5File(mesh.mpi_comm(),
                "results/h5-ernie-parcellation.h5", "w")
hdf5.write(mesh, "/mesh")
hdf5.write(regions, "/regions")
hdf5.close()
```

The result can be seen alongside the original parcellation in Figure 4.6.

4.4.2 Mapping parcellations respecting subdomains

By careful inspection of the mesh-based representation of the parcellation in Figure 4.6 (right), we find some artifacts in the labeling of the cells. Indeed, since the finite element mesh is not aligned with the segmentation, the previous direct mapping between cell midpoints and voxels can lead to minor inaccuracies. In this section, we therefore focus on improving our mesh-based parcellation representation by using subdomain information.

Specifically, we will show you how to:

- read in a mesh with subdomain information. The mesh may contain many subdomains, such as those defined by a regional parcellation label. The mesh subdomain structure may also be simple, such as the gray/white matter subdomain regions discussed in Section 4.1.2;
- for each cell in every subdomain, determine if we should alter the label assigned to the cell by inspecting the label of the neighboring cells;
- store the subdomain and parcellation information together.

Converting meshes and mesh data between different formats

In Chapter 4.1, we created a mesh with gray and white matter labels stored as subdomain information in the .mesh format. Our next step is to convert this information to a FEniCS-compatible format. We will write a convenient Python script for this common operation and use the opportunity to illustrate the use of an **ArgumentParser** to read in arguments from the command line, instead of coding the arguments directly into the script (included in mri2fem/chp4/convert_to_dolfin_mesh.py).

In this script, we will use the **argparse** package to enable us to pass in command-line arguments that inform the conversion process.

```
import argparse

def write_mesh_to_xdmf(meshfile, xdmfdir):
```

Our argument parser set-up looks like this:

```
if __name__ =='__main__':
    parser = argparse.ArgumentParser()
    parser.add_argument('--meshfile', type=str)
    parser.add_argument('--hdf5file', type=str)
    parser.add_argument('--xdmfdir', type=str,
                        default="tmp")
    Z = parser.parse_args()
```

and we then call two functions (described below):

```
    write_mesh_to_xdmf(Z.meshfile, Z.xdmfdir)
    write_xdmf_to_h5(Z.xdmfdir, Z.hdf5file)
```

The script can then be called as, for example:

```
$ cd mri2fem/chp4
$ python convert_to_dolfin_mesh.py --meshfile ernie-gw.mesh
--hdf5file ernie-gw.h5
```

We use meshio to first read the .mesh file, its data associated with the mesh cells (subdomains), and its data associated with mesh facets (boundaries). We then write these in the FEniCS-readable .xdmf format as separate files.

```
# Read the .mesh file into meshio
mesh = meshio.read(meshfile)

# Extract subdomains and boundaries between regions
# into appropriate containers
points = mesh.points
tetra    = {"tetra": mesh.cells_dict["tetra"]}
triangles = {"triangle": mesh.cells_dict["triangle"]}
subdomains = {"subdomains": [mesh.cell_data_dict["medit:
                                   ref"]["tetra"]]}
boundaries = {"boundaries": [mesh.cell_data_dict["medit:
                                   ref"]["triangle"]]}

# Write the mesh to xdmfdir/mesh.xdmf
xdmf = meshio.Mesh(points, tetra)
meshio.write("%s/mesh.xdmf" % xdmfdir, xdmf)

# Write the subdomains of the mesh
xdmf = meshio.Mesh(points, tetra, cell_data=subdomains)
meshio.write("%s/subdomains.xdmf" % xdmfdir, xdmf)

# Write the boundaries/interfaces of the mesh
xdmf = meshio.Mesh(points, triangles, cell_data=boundaries)
meshio.write("%s/boundaries.xdmf" % xdmfdir, xdmf)
```

Subsequently, FEniCS can read the files mesh.xdmf and subdomains.xdmf, created above, into a Mesh object and MeshFunction object, respectively. FEniCS can then write the data from both objects into a single .h5 file:

```
import dolfin as df
# Read .xdmf mesh into a FEniCS Mesh
mesh = df.Mesh()
with df.XDMFFile("%s/mesh.xdmf" % xdmfdir) as infile:
    infile.read(mesh)

# Read cell data to a MeshFunction (of dim n)
n = mesh.topology().dim()
```

```
subdomains = df.MeshFunction("size_t", mesh, n)
with df.XDMFFile("%s/subdomains.xdmf" % xdmfdir) as infile
                              :
    infile.read(subdomains, "subdomains")

# Read facet data to a MeshFunction (of dim n-1)
boundaries = df.MeshFunction("size_t", mesh, n-1, 0)
with df.XDMFFile("%s/boundaries.xdmf" % xdmfdir) as infile
                              :
    infile.read(boundaries, "boundaries")

# Write all files into a single h5 file.
hdf = df.HDF5File(mesh.mpi_comm(), hdf5file, "w")
hdf.write(mesh, "/mesh")
hdf.write(subdomains, "/subdomains")
hdf.write(boundaries, "/boundaries")
hdf.close()
```

The mesh, subdomain, and boundary information saved above can be read back into FEniCS as follows (with the filename given in `hdf5file`):

```
# Read the mesh and mesh data from .h5:
mesh = Mesh()
hdf = HDF5File(mesh.mpi_comm(), hdf5file, "r")
hdf.read(mesh, "/mesh", False)

d = mesh.topology().dim()
subdomains = MeshFunction("size_t", mesh, d)
hdf.read(subdomains, "/subdomains")
boundaries = MeshFunction("size_t", mesh, d-1)
hdf.read(boundaries, "/boundaries")
hdf.close()
```

Masking data and improved parcellation identification

Let us now assume that we have available the `mesh`, `subdomains`, as well as the image `data` from the parcellation file (e.g `wmparc.mgz`) and the RAS to voxel space transform `ras2vox`, as described in Chapter 4.4.1. Our next steps are to:

- find the RAS coordinates of all cells (midpoints), and map these to voxel space indices and,
- map parcellation values to mesh cells in a manner that ensures that the parcellation regions do not extend across subdomains.

In particular, for each mesh cell, we will now examine a neighborhood of the corresponding voxel values to pick the most frequent adjacent one as the matched parcellation value.

We first extract all the RAS coordinates of the cell midpoints,

```
# Extract RAS coordinates of cell midpoints
xyz = numpy.array([cell.midpoint()[:]
                      for cell in cells(mesh)])
```

before converting these to voxel coordinates and indices, as before:

```
# Convert to voxel space and voxel indices: for cell c,
# i[c], j[c], k[c] give the corresponding voxel indices.
abc = apply_affine(ras2vox, xyz).T
ijk = numpy.rint(abc).astype("int")
(i, j, k) = ijk
```

Now, we create two arrays. The first array, vox2sub, provides a map from the index of a voxel to its subdomain tag; the second array, regions, will hold the parcellation tags.

```
# Create a map from voxel index to subdomain tag
# Note use of NumPy's "fancy" indexing:
vox2sub = numpy.zeros(data.shape)
vox2sub[i, j, k] = subdomains.array()

# Create new array for the parcellation tags:
N = mesh.num_cells()
regions = numpy.zeros(N)
```

Now we adjust tags by a consensus method. We do this by applying the following algorithm: for a fixed subdomain, iterate over the cells in that subdomain; for each cell, examine the voxel data (region label) associated to the neighbors of that cell but only those neighbors which also live in the same subdomain; set the region of the cell to the most common region of its neighbors. This process is repeated for each subdomain in the mesh; the algorithm is implemented below:

```
# Extract unique mesh subdomain tags,
# and iterate over these:
subdomain_tags = numpy.unique(subdomains.array())
for tag in subdomain_tags:
    # Zero out voxel data not associated with the current
    # subdomain
    masked_data = (vox2sub == tag)*data
```

```
        # Iterate of all cells in this subdomain
        for c in range(N):
            if (subdomains[c] == tag):
                # Find and set the most common (non-zero)
                # adjacent parcellation tag
                regions[c] = adjacent_tag(masked_data,
                                          i[c], j[c], k[c])
```

We can then update the subdomain array

```
    # Update the subdomains array with the parcellation tags
    if not specific_tags:
        subdomains.array()[:] = regions
    else:
        for tag in specific_tags:
            subdomains.array()[regions == tag] = tag
```

and store the resulting mesh data again:

```
    # Now store everything to a new file
    hdf = HDF5File(mesh.mpi_comm(), out_hdf5, "w")
    hdf.write(mesh, "/mesh")
    hdf.write(subdomains, "/subdomains")
    hdf.write(boundaries, "/boundaries")
    hdf.close()
```

The `adjacent_tag` function reads

```
def adjacent_tag(data, i, j, k, Mmin=3, Mmax=10):
    # Given an image voxel index (i, j, k), examine the
    # image data in the voxel neighborhood, and identify
    # the most common non-zero value among these. Start at
    # a neighborhood of radius Mmin, and increase if needed.
    for m in range(Mmin, Mmax):

        # Extract the data values from the neighborhood
        values = data[i-m:i+m+1, j-m:j+m+1, k-m:k+m+1]

        # Reshape values from (2m+1, 2m+1, 2m+1) to list:
        v = values.reshape(1, -1)

        # Identify unique non-zero (positive) values and
        # the number of each
        pairs, counts = numpy.unique(v[v > 0],
                                     return_counts=True)

        # Return the most common non-zero tag:
        success = counts.size > 0
```

```
    if success:
        return pairs[counts.argmax()]

  return 0
```

The complete script is included in `mri2fem/chp4/add_parcellations.py`
and can be tested, for example, via

```
$ python3 convert_to_dolfin_mesh.py --meshfile
ernie-brain-32.mesh --hdf5file ernie-brain-32.h5
$ python3 add_parcellations.py --in_hdf5
ernie-brain-32.h5 --in_parc wmparc.mgz --out_hdf5
results/ernie-brain-subdomains-tags.h5 --add 17 1028 1035 3028
3035
```

4.5 Refinement of parcellated meshes

We end this chapter by examining how we can refine the generated meshes
and, in particular, how we can refine local regions. FEniCS supports global
and local mesh refinement and adaptivity through the functions **refine** and
adapt. The latter (**adapt**) is particularly useful for refining mesh functions
and associated mesh data, in addition to the refinement of the mesh itself. The code snippets presented here are included in a complete context
in `mri2fem/chp4/refine_mesh_tags.py`.

4.5.1 Extending the Python interface of DOLFIN/FEniCS

DOLFIN, the problem-solving interface to FEniCS, provides a C++ and a
Python interface. The Python interface is generated from the C++ interface
using pybind11, but the entire C++ API is not exposed. Instead, a user can
generate their own Python bindings fairly easily. Since parts of the **adapt**
interface are not, by default, available in Python, we illustrate how to use
pybind11 to compile our own FEniCS wrappers here.

We provide Python bindings for the **adapt** function as follows:

```
cpp_code = """
#include<pybind11/pybind11.h>
#include<dolfin/adaptivity/adapt.h>
#include<dolfin/mesh/Mesh.h>
#include<dolfin/mesh/MeshFunction.h>

namespace py = pybind11;

PYBIND11_MODULE(SIGNATURE, m) {
  m.def("adapt", (std::shared_ptr<dolfin::MeshFunction<std::
                             size_t>> (*)(const dolfin::
                             MeshFunction<std::size_t>&,
                             std::shared_ptr<const dolfin
                             ::Mesh>)) &dolfin::adapt, py
                             ::arg("mesh_function"), py::
                             arg("adapted_mesh"));
  m.def("adapt", (std::shared_ptr<dolfin::Mesh> (*)(const
                             dolfin::Mesh&)) &dolfin::
                             adapt );
  m.def("adapt", (std::shared_ptr<dolfin::Mesh> (*)(const
                             dolfin::Mesh&,const dolfin::
                             MeshFunction<bool>&)) &dolfin
                             ::adapt );
}
"""
adapt = compile_cpp_code(cpp_code).adapt
```

We also notice that this function is overloaded, and we provide bindings
to three versions with different signatures. After the code is executed, the
bindings are compiled by the Python module. However, note that the module
will not be added to the DOLFIN library, but resides in the user's local cache.

4.5.2 Refining certain regions of parcellated meshes

It may be desirable to refine a mesh - either globally or within particular
regions. In this section, we discuss two options for refinement; global and local.
The code snippets in this section can be found in the file refine_mesh_tags.
py. Let us assume that in Python we have a FEniCS mesh with FEniCS mesh
functions subdomains and boundaries, for instance, extracted from an .h5
file, as illustrated in Chapter 4.4.2.

One option is, then, to refine the mesh globally, meaning that we refine all the cells of the mesh. We would then also like to refine or adapt the associated mesh functions to the refined mesh. We can accomplish this as follows:

```
# Initialize connections between all mesh entities, and
# use a refinement algorithm that remember parent facets
mesh.init()
parameters["refinement_algorithm"] = \
    "plaza_with_parent_facets"

# Refine globally if no tags given
if not tags:
    # Refine all cells in the mesh
    new_mesh = adapt(mesh)

    # Update the subdomain and boundary markers
    adapted_subdomains = adapt(subdomains, new_mesh)
    adapted_boundaries = adapt(boundaries, new_mesh)
```

Notice that the script `refine_mesh_tags.py` refines the mesh globally if the `tags` is empty. The `tags` list can be specified, or not, as an input argument to `refine_mesh_tags.py`. For global refinement, we do not want to specify tags; the following command will globally refine the mesh, `ernie-brain-32.h5`, created in the previous section:

```
$ python3 refine_mesh_tags.py --in_hdf5 ernie-brain-32.h5
--out_hdf5 ernie-brain-32-refined.h5
```

Alternatively, we can refine the mesh locally. In the code, this is done by providing the `adapt` function with a Boolean cell-based mesh function that sets a value of `True` or `False` at each mesh cell (i.e. tetrahedron). A value of `True` indicates that the cell should be refined, while `False` indicates it should not. The code starts by setting every cell's associated value to `False`. Then, we loop over the mesh cells and check to see if the cell's region matches a region in the `tags` list; if so, we mark this cell for refinement by changing its associated value to `True`. The relevant code snippet is:

```
else:
    # Create markers for local refinement
    markers = MeshFunction("bool", mesh, d, False)

    # Iterate over given tags, label all cells
    # with this subdomain tag for refinement:
    for tag in tags:
        markers.array()[subdomains.array()==tag] = True
```

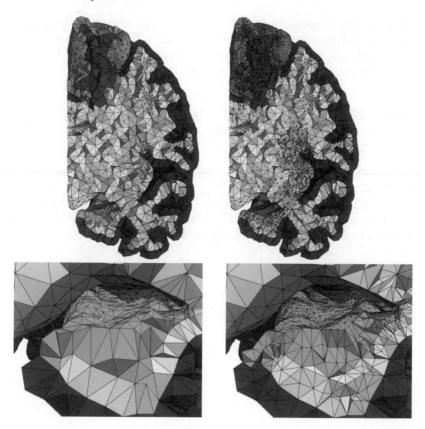

Fig. 4.7 Illustration of a locally refined left hemisphere mesh. The left figures show the original mesh with labeled (color coded) regions. The right figures show the locally refined mesh. Only particular regions have been refined; compare the green region to the red region. Local refinement is carried out using `refine_mesh_tags.py` and using the `--refine_tag` option.

```
# Refine mesh according to the markers
new_mesh = adapt(mesh, markers)

# Update subdomain and boundary markers
adapted_subdomains = adapt(subdomains, new_mesh)
adapted_boundaries = adapt(boundaries, new_mesh)
```

In practice, the `tags` list is populated by passing in command line arguments to the script. For instance, the following command will refine the cells in regions 17 (left hippocampus), 1028 (left superior frontal cortex), 1035 (left insular cortex), 3028 (left superior frontal white matter) and 3035 (left insular white matter).

```
$ python3 refine_mesh_tags.py --in_hdf5 ernie-brain-32.h5
--out_hdf5 ernie-brain-32-refine-tags.h5 --refine_tag 17 1028
1035 3028 3035
```

We remind the reader that the various tags can be found by opening Freeview (c.f. Chapter 2.4), selecting $\boxed{\text{File}\rightarrow\text{Load Volume}}$ from the top menu, selecting `mri2fem/chp4/wmparc.mgz`, and then setting the `Color map` option (in the left pane) to `Lookup Table`. The local refinement of a mesh with labeled parcellation regions is illustrated in Figure 4.7.

Chapter 5
Introducing directionality with diffusion tensors

In this chapter, we focus on how to transfer information from diffusion tensor imaging (DTI) data to our finite element methods. To do so, we will need to overcome a few practical challenges. In particular, the raw DTI data can contain non-physiological data, especially near the CSF. Moreover, the raw DTI data is represented both in terms of a different coordinate system and at a different resolution than the computational mesh. To overcome the first challenge, we will use local extrapolation of nearby valid values; to overcome the second challenge, we will co-register [1] the data with the images used to construct the computational mesh.

Specifically, we will:

- process the diffusion tensor images to extract mean diffusivity and fractional anisotropy [2] data, and
- map the DTI tensor data into a finite element representation created from the T1-weighted images.

---

[1] See Section 5.2.3.

[2] Mean diffusivity and fractional anisotropy are defined in (5.1) and (5.2), respectively.

5.1 Extracting mean diffusivity and fractional anisotropy

5.1.1 Extracting and converting DTI data

The DTI data must first be extracted from a DICOM dataset. We use Dicom-Browser to extract a DTI series from the book data-set in Chapter 2.3, and the resulting files are available in `dicom/ernie/DTI`. Our next task is to convert the extracted DTI images to a single volume image and to produce supplementary information files about the DTI image data for downstream postprocessing. Various open source tools are available for the processing of DTI data [60]. Here, we continue to use FreeSurfer and its associated command-line tools. As in chapter 3.1.2, we can select any of the files extracted from the DICOM DTI data (`dicom/ernie/DTI`) to start the process; here, we arbitrarily choose IM_1496 and launch the FreeSurfer command `mri_convert`:

```
$ cd dicom/ernie/DTI
$ mri_convert IM_1496 dti.mgz
```

This process, when successful, creates three files: `dti.mgz`, `dti.bvals`, and `dti.voxel_space.bvec`. The last two, plain text files, contain information regarding the b-values and b-vectors associated with the DTI data. The b-vectors and b-values are selected as part of the imaging process; they determine the direction (b-vector) and strength (b-value) of the pulsed magnetic diffusion gradient used during the diffusion weighted imaging scan. For instance, Figure 5.1 shows an axial slice measured with the same choice of b-value but different b-vectors. Once the scan has taken place, we can read this information but it cannot be altered without scanning the patient again.

5.1.2 DTI reconstruction with FreeSurfer

Next, we aim to reconstruct comprehensive DTI data from the volume, b-value, and b-vector files using the FreeSurfer command `dt_recon`. The command takes an input volume (following `--i`), b-vector and b-values files (following `--b`), an output directory `--o`, and the `recon-all` subject ID `--s` (see Chapter 3.1.2). Within our book data directory `dicom/ernie/DTI`, we can launch the following commands:

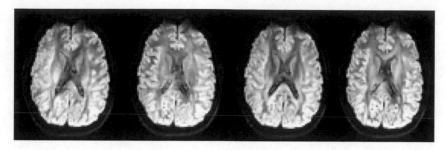

Fig. 5.1 Axial DTI slices measured with different b-vectors. The resolution in the diffusion tensor image is typically lower (here, 96x96x50) compared to that in the T1 images; the latter are, canonically, 256x256x256.

```
$ export SUBJECTS_DIR=my-freesurfer-dir
$ dt_recon --i dti.mgz --b dti.bvals dti.voxel_space.bvecs --s
ernie --o $SUBJECTS_DIR/ernie/dti
```

with `my-freesurfer-dir` replaced by the FreeSurfer subject's directory (e.g. `freesurfer/` from the book data-set).

This command produces multiple output files, [3] including `tensor.nii.gz`, `register.dat`, and `register.lta`. The registration in `dt_recon` uses the registration command `bbregister` [4] to register the DTI data [3]. Files with the suffix .nii are in the NIfTI format. Of these, `tensor.nii.gz` is the spatially varying diffusion tensor. Further, an eigendecomposition of this tensor in terms of spatially varying eigenvalues λ_1, λ_2, and λ_3 and eigenvectors v_1, v_2, and v_3 is given in the files `eigvals.nii.gz` and `eigvec1.nii.gz`,`eigvec2.nii.gz`, and `eigvec3.nii.gz`.

---

[3] The command above will store the files in $SUBJECTS_DIR/ernie/dti. Alternatively, you can run `mri2fem/chp5/all.sh` which will create a directory `mri2fem/chp5/ernie-dti` that includes the same set of the files as well. You will need FSL installed to use `dt_recon` (see Chapter 2.4.1).

[4] This registration step is done automatically by FreeSurfer using the subject's previously FreeSurfer-processed data that is assumed to be available at this stage of the book; see Chapter 3.1.2 for the necessary steps. The mathematical details of co-registration are further discussed in Section 5.2.3.

5.1.3 Mean diffusivity and fractional anisotropy

In addition, dt_recon produces the NIfTI files adc.nii.gz and fa.nii.gz for the mean (or apparent) diffusivity (MD) and fractional anisotropy (FA), respectively. The mean diffusivity is given by

$$\text{MD} = \frac{1}{3}(\lambda_1 + \lambda_2 + \lambda_3), \tag{5.1}$$

and fractional anisotropy is defined [39] by

$$\text{FA}^2 = \frac{1}{2} \frac{(\lambda_1 - \lambda_2)^2 + (\lambda_2 - \lambda_3)^2 + (\lambda_3 - \lambda_1)^2}{\lambda_1^2 + \lambda_2^2 + \lambda_3^2}. \tag{5.2}$$

NIfTI files can be viewed in ParaView. You might first need to enable

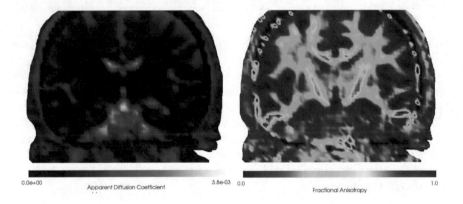

0.0e+00   Apparent Diffusion Coefficient   3.8e-03 0.0   Fractional Anisotropy   1.0

Fig. 5.2 Mean diffusivity (left) and fractional anisotropy (right) as shown in ParaView.

the NIfTI viewer plugin by selecting the ParaView menu option labeled [Tools→Manage Plugins], selecting [5] [AnalyzeNifTIReaderWriter] and then clicking [Load Selected]. You can then open and view .nii files in ParaView,

---

[5] The correct plugin may also be named AnalyzeNifTIIO in earlier versions (i.e. before 5.7.0) of ParaView.

just as you would any other file. To unzip .nii.gz to .nii, one can use `mri_convert`:

```
$ mri_convert adc.nii.gz adc.nii
$ mri_convert fa.nii.gz fa.nii
```

Let us open `adc.nii` and verify that we can reproduce Figure 5.2 (left); the process will be the same for `fa.nii`. After loading the AnalyzeNifTIIO plugin, described above, and opening `adc.nii` click Apply . You will likely see an empty three-dimensional cube in the view window. In the left pane, find the `Representation` option and change this to `Volume`. You should now see something that looks similar to a 'brain in a box' viewed from the top. Now click Filters→Alphabetical and select `Slice`. In the left pane, find the option labeled `Normal`; it should be under the option labeled `Origin`. Change the `Normal` from 1 0 0 to 0 1 0 and click Apply .

In the left pane, once more, hide the object `adc.nii` by clicking the picture of the eye next to its name. Now, rotate the view window so that you can see the X–Z plane; the result should look similar to Figure 5.2. We can make it look more similar by changing the color scheme. In the left pane, find the section labeled `Coloring`. Mouse over the buttons here until you find the button labeled `Choose preset`. Click this and select the `Black, blue and white` color scheme, click Apply and then close the color scheme preset window. The image you see now was saved and post-processed to remove the border outside the skull to produce Figure 5.2 (left). You can repeat these steps with `fa.nii`, this time using the `jet` color scheme, to reproduce Figure 5.2 (right).

The average FA value is generally around 0.5 and changes by around 2% between day and night [68]. Anisotropy decreases with age, declining around 14% between 30 to 80 years [40] and can change by up to 50% in certain areas of the brain of a person with Alzheimer's disease compared with healthy subjects [49]. In the `ernie` data (Figure 5.2), the median white matter FA value is 0.3, with a minimum of 0.009 and a maximum of 0.9998.

5.2 Finite element representation of the diffusion tensor

In this section, we:

- ensure that the DTI data have a valid eigendecomposition (with positive eigenvalues),

- map the DTI tensor into a finite element tensor function defined on a finite element mesh, and
- briefly discuss co-registration.

5.2.1 Preprocessing the diffusion tensor data

The DTI data can be quite rough compared to the T1 data and our corresponding finite element meshes; DTI data is typically at a low resolution of 96x96x50
while T1 resolution is typically much higher at 256x256x256. [6] Moreover, the
signal can be disturbed near the cerebrospinal fluid (CSF), which makes the
data in certain areas of the cortical gray matter and in regions near the ventricle system less reliable. Indeed, inspection of the eigenvalues of the DTI tensor
shows non-physiological (zero and/or negative) eigenvalues. To ensure a phys-
iologically (and mathematically) reasonable diffusion tensor, we recommend
preprocessing the diffusion tensor prior to numerical simulation. In particular,
in this chapter we present two scripts that:

- check the DTI tensor data for non-physiological values and
- replace non-physiological with physiological values in the DTI tensor,

respectively.

Creating brain masks

First, we will use FreeSurfer to create masks of the brain. A mask is a type
of filter where voxels (significantly) outside the brain are set to zero and all
other voxels are set to a value of one. Using our white matter parcellation data
(included in `freesurfer/ernie/mri/wmparc.mgz`), we can create brain masks
as follows:

```
$ mri_binarize --i wmparc.mgz --gm --dilate 2 --o mask.mgz
```

The `dilate` flag determines the extent to which the mask should be extended
outside the brain surface provided by `wmparc.mgz`. Examples of such masks
are shown in Figure 5.3.

---

[6] See Figure 2.3 (left) versus Figure 5.1.

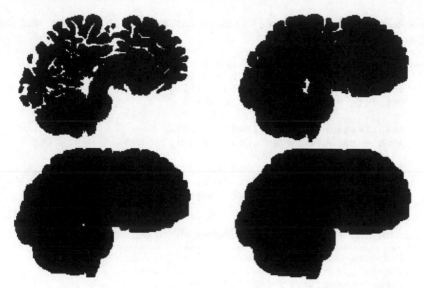

Fig. 5.3 Brain masks created using mri_binarize with dilate ranging from zero to three.

Examining the DTI data values

We can work with the DTI data in a very similar manner as we did for the parcellation (image) data in Chapter 4.4.1. We will again use NiBabel to load the image data, use the **vox2ras** functions for the mapping between the different image coordinate systems (DTI voxel space and T1 voxel space), and process the data as NumPy arrays. The complete script can be run as

```
$ cd mri2fem/chp5
$ python3 check_dti.py --dti tensor.nii.gz --mask mask.mgz
```

We import the key packages:

```
import argparse
import numpy
import nibabel

from nibabel.processing import resample_from_to
numpy.seterr(divide='ignore', invalid='ignore')
```

We define the function `check_dti_data` that takes the DTI tensor and mask files as input:

```python
def check_dti_data(dti_file, mask_file, order=0):
    # Load the DTI image data and mask:
    dti_image = nibabel.load(dti_file)
    dti_data = dti_image.get_fdata()

    mask_image = nibabel.load(mask_file)
    mask = mask_image.get_fdata().astype(bool)

    # Examine the differences in shape
    print("dti shape  ", dti_data.shape)
    print("mask shape ", mask.shape)
    M1, M2, M3 = mask.shape
```

Now, the important coordinate transformations can be handled as follows:

```python
    # Create an empty image as a helper for mapping
    # from DTI voxel space to T1 voxel space:
    shape = numpy.zeros((M1, M2, M3, 9))
    vox2ras = mask_image.header.get_vox2ras()
    Nii = nibabel.nifti1.Nifti1Image
    helper = Nii(shape, vox2ras)

    # Resample the DTI data in the T1 voxel space:
    image = resample_from_to(dti_image, helper, order=order)
    D = image.get_fdata()
```

Before computing eigenvalues, we run

```python
    # Reshape D from M1 x M2 x M3 x 9 into a N x 3 x 3:
    D = D.reshape(-1, 3, 3)

    # Compute eigenvalues and eigenvectors
    lmbdas, v = numpy.linalg.eigh(D)
```

and we compute the fractional anisotropy and check the validity of each voxel value, as follows:

```python
    # Compute fractional anisotropy (FA)
    FA = compute_FA(lmbdas)

    # Define valid entries as those where all eigenvalues are
    # positive and FA is between 0 and 1
    positives = (lmbdas[:,0]>0)*(lmbdas[:,1]>0)*(lmbdas[:,2]>0)
    valid = positives*(FA < 1.0)*(FA > 0.0)
    valid = valid.reshape((M1, M2, M3))
```

```
# Find all voxels with invalid tensors within the mask
ii, jj, kk = numpy.where((~valid)*mask)
print("Number of invalid tensor voxels within the mask ROI:
                        ", len(ii))

# Reshape D from N x 3 x 3 to M1 x M2 x M3 x 9
D = D.reshape((M1,M2,M3,9))

return valid, mask, D
```

The above snippet makes use of the function compute_FA, which is also defined in check_dti.py, to compute (5.2). The result is a vector FA whose entries contain the fractional anisotropy computed at each available DTI data location. The term positives is a binary vector, with the same number of entries as FA; it has a value of one if all three of the eigenvalues for the region corresponding to the array index are positive, and zero otherwise. The vector valid is therefore a second binary vector whose indices correspond to the locations where DTI data are available. The value at each index of valid is one precisely when all of the eigenvalues are positive and the fractional anisotropy there is larger than zero but less than one. The valid vector is therefore a mask that indicates where the DTI tensor contains physically admissible values. The valid mask is then reshaped [7] to fit the dimensions of the original mask, created from the mask_file, and the number of zeros, corresponding to invalid entries, is computed and reported in the final lines.

Improving DTI values by extrapolation and resampling to T1 space

If numerous invalid DTI voxel data are reported, by check_dti.py as discussed above, we can attempt to improve the DTI data by extrapolating from adjacent valid voxel locations to correct nearby invalid data. The correction script is mri2fem/chp5/clean_dti_data.py and can be run as

```
$ cd mri2fem/chp5
$ python3 clean_dti_data.py --dti tensor.nii.gz --mask mask.mgz
--out tensor-clean.nii
```

---

[7] The term *reshaped* here means that the (tensor) data is reorganized into an expected form. An example would be reshaping a 1×9 (row) tensor to a 3×3 (matrix) tensor by putting the first three entries of the 1×9 tensor in the first row, the next three in the second row and the last three in the final row of the 3×3 tensor.

and the main functionality reads

```
def clean_dti_data(dti_file, mask_file, out_file, order=3,
                   max_search=9):
    valid, mask, D = check_dti_data(dti_file, mask_file,
                                    order=order)
    # Zero out "invalid" tensor entries outside mask,
    # and extrapolate from valid neighbors
    D[~mask] = numpy.zeros(9)
    D[(~valid)*mask] = numpy.zeros(9)
    ii, jj, kk = numpy.where((~valid)*mask)
    for i, j, k in zip(ii, jj, kk):
        D[i, j, k, :] = \
            find_valid_adjacent_tensor(D, i, j, k, max_search)

    # Create and save clean DTI image in T1 voxel space:
    mask_image = nibabel.load(mask_file)
    M1, M2, M3 = mask.shape
    shape = numpy.zeros((M1, M2, M3, 9))

    vox2ras = mask_image.header.get_vox2ras()
    Nii = nibabel.nifti1.Nifti1Image
    dti_image = Nii(D, vox2ras)

    nibabel.save(dti_image, out_file)
```

The first operation carried out by the `clean_dti_data` function is to call `check_dti_data`, which we discussed in the previous section. Recall that, among other things, the `check_dti_data` function returns a tensor representation (D) of the DTI data that has been converted from DTI voxel space coordinates to T1 voxel space coordinates. [8] Next, we will search for a valid tensor in directly adjacent voxels using the function `find_valid_adjacent_tensor` defined in `clean_dti_data.py`. If no valid tensor is found nearby, the search range is iteratively increased.

The script determines that a nearby tensor, in the valid region, contains valid data if a non-zero mean diffusivity (MD) is also calculated there. Once one or more, nearby valid value(s) are found, replacement data is chosen. If only one valid value is found, it is directly used. If there are multiple valid

---

[8] T1 coordinates are the same coordinates used by the computational meshes that were constructed, in previous chapters, from the surfaces extracted from FreeSurfer segmented T1 data. Thus, D is now expressed in terms of coordinates that make sense when used alongside the computational meshes

tensors within the search range, the tensor data [9] with MD value closest to the median of the non-zero MD is chosen as a replacement:

```python
def find_valid_adjacent_tensor(data, i, j, k ,max_iter):
    # Start at 1, since 0 is an invalid tensor
    for m in range(1, max_iter+1) :
        # Extract the adjacent data to voxel i, j, k
        # and compute the mean diffusivity.
        A = data[i-m:i+m+1, j-m:j+m+1, k-m:k+m+1,:]
        A = A.reshape(-1, 9)
        MD = (A[:, 0]+ A[:, 4] + A[:,8])/3.

        # If valid tensor is found:
        if MD.sum() > 0.0:
            # Find index of the median valid tensor, and return
            # corresponding tensor.
            index = (numpy.abs(MD - numpy.median(MD[MD>0]))).
                                                          argmin()
            return A[index]

    print("Failed to find valid tensor")
    return data[i, j, k]
```

5.2.2 Representing the DTI tensor in FEniCS

With the DTI data checked and potentially improved, we are now ready to map our preprocessed DTI image (now in T1 voxel space) onto a FEniCS mesh. We will use the code located in mri2fem/chp5/dti_data_to_mesh.py to accomplish this task. To begin, we assume that we have a mesh available (e.g. ernie-brain-32.h5 from Chapter 4.4.2), that we have loaded the clean DTI image and data in dti_image and dti_data, respectively, and that we have the ras2vox transform associated with this image. We can retrieve the vox2ras and ras2vox transformations associated with the data by

---

[9] Because of the way the valid mask is constructed, a tensor with invalid data can violate either the required condition that all of the eigenvalues must satisfy $\lambda_i > 0$ or the required condition that the FA must satisfy $0 <$ FA < 1. In either case, the search for a nearby valid tensor identifies a nearby candidate and replaces the whole of the tensor information at the invalid tensor location. Thus, all of required conditions are satisfied, at the previously invalid location, after the data replacement.

```
# Transformation to voxel space from mesh coordinates
vox2ras = dti_image.header.get_vox2ras_tkr()
ras2vox = numpy.linalg.inv(vox2ras)
```

To represent the diffusion tensor in FEniCS, we create a FEniCS `Function` over a `TensorFunctionSpace` of (discontinuous) piecewise constant polynomial fields (`"DG"`, 0):

```
# Create a FEniCS tensor field:
DG09 = TensorFunctionSpace(mesh, "DG", 0)
D = Function(DG09)
```

For each cell, we need to associate an identifying coordinate value so that we can associate the cells of our mesh to the voxel data. One possibility is to extract the cell midpoints as we have done before; here, we opt to extract the coordinates of the degrees of freedom associated with a DG `FunctionSpace` object that we will define on our mesh and convert these to voxel indices:

```
# Get the coordinates xyz of each degree of freedom
DG0 = FunctionSpace(mesh, "DG", 0)
imap = DG0.dofmap().index_map()
num_dofs_local = (imap.local_range()[1] \
                  - imap.local_range()[0])
xyz = DG0.tabulate_dof_coordinates()
xyz = xyz.reshape((num_dofs_local, -1))

# Convert to voxel space and round off to find
# voxel indices
ijk = apply_affine(ras2vox, xyz).T
i, j, k = numpy.rint(ijk).astype('int')
```

The above snippet first retrieves the coordinates of the `TensorFunctionSpace` degrees of freedom on our mesh and applies the `ras2vox` transformation to determine coordinates in voxel space.

We can now reshape the DTI data into a cell-wise structure based on the extracted indices[10] (now in voxel space):

```
# Create a matrix from the DTI representation
D1 = dti_data[i, j, k]
```

[10] Voxels are located based on the degree of freedom (DOF) coordinates from the `FunctionSpace` object. This approach guarantees that there are no missing values as every coordinate maps to some voxel. However, some voxels may correspond to more than one mesh cell as there may be more cells in the mesh than there are voxels e.g. if the mesh has a lower resolution than the resolution of the T1 (voxel) image space.

```
print(D1.shape)
```

With the reshaped DTI data in hand, we assign these to a FEniCS tensor field, D, allowing the data to be saved alongside the mesh data.

```
# Assign the output to the tensor function
D.vector()[:] = D1.reshape(-1)
```

The FEniCS tensor field DTI data can be saved alongside the mesh for later use in FEniCS simulations with

```
# Now store everything to a new file - ready for use!
hdf = HDF5File(mesh.mpi_comm(), outfile, 'w')
hdf.write(mesh,"/mesh")
hdf.write(D, "/DTI")
```

The resulting fiber directions, shown in Figure 5.4, can be inspected visually.

5.2.3 A note on co-registering DTI and T1 data

As we have seen, FreeSurfer uses several different coordinate systems to label the position of data in its various output files. Thus, to combine different types of data into something we can use in FEniCS simulations, we need to extract information about the different coordinate systems used in the files and be able to map between these different coordinate systems. This process is known as co-registration. The scripts we have presented use NiBabel functionality to handle co-registration; this section provides additional information regarding co-registration, for both context and completeness.

In short, let $x_1 = (x_1, y_1, z_1)$ and $x_2 = (x_2, y_2, z_2)$ represent the same physiological point in $\mathbb{R}^3$ but represented with respect to two different coordinate systems (bases). Then, there is an affine transformation such that

$$x_2 = A x_1 + b, \tag{5.3}$$

for $A \in \mathbb{R}^{3 \times 3}$ and $b \in \mathbb{R}^3$. The mapping is often stored instead as a 4×4 matrix, where the last row can be ignored. As this equivalent 4×4 representation often appears in the discussions, and software documentation, within the neuroimaging community, we also show it here; the above affine transformation (5.3) can also be written as:

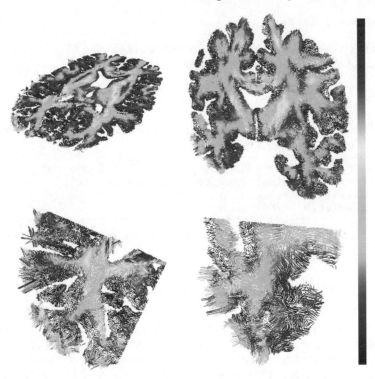

Fig. 5.4 Upper panels show fiber directions (DTI eigenvectors) colored by the fractional anisotropy in the axial and coronal planes. The lower panels show a zoom focusing on the boundary between gray matter and the cerebrospinal fluid. Note that the vector nature of the data can be seen more clearly in bottom panel images where the fibers can be seen to have clear directionality.

$$
\begin{bmatrix} x_2 \\ y_2 \\ z_2 \\ 1 \end{bmatrix} = \begin{bmatrix} a_{11} & a_{12} & a_{13} & b_1 \\ a_{21} & a_{22} & a_{23} & b_2 \\ a_{31} & a_{32} & a_{33} & b_3 \\ 0 & 0 & 0 & 1 \end{bmatrix} \begin{bmatrix} x_1 \\ y_1 \\ z_1 \\ 1 \end{bmatrix},
$$

where the a_{ij} are the entries of the matrix A and the b_i are the entries of the vector b.

The term *co-registration* specifically refers to the determination of the transformation matrix A and vector b corresponding to a pair of files. A key step in the co-registration of T1 and DTI images, or any pair of images in general, is to ascertain the type of coordinate system used when initially storing these images. Towards this end, we can make use of the `mri_info` command. Coordinate system information regarding the FreeSurfer-processed T1 images is stored in the file `orig.mgz`. We can interrogate this file by:

```
$ cd $SUBJECTS_DIR/ernie/mri
$ mri_info orig.mgz --orientation
LIA
```

The output **LIA** means that the T1 image files were generated with respect to the 'Left Inferior Anterior' coordinate system (see [3] for details). Coordinate system information regarding the FreeSurfer-processed DTI images is stored in the file `tensor.nii.gz`. We can interrogate this file, once more using `mri_info`, by:

```
$ cd $SUBJECTS_DIR/ernie/dti
$ mri_info tensor.nii.mgz --orientation
LPS
```

The coordinate systems can be understood as follows: the positive direction in the sagittal plane can be either (L)eft or (R)ight, the positive direction in the coronal plane can be either (P)osterior or (A)nterior, and the positive direction in the axial plane can be either (I)nferior or (S)uperior. Furthermore, the order of the planes can be different, that is, the third axis might not correspond to the axial plane. For instance, let us examine the coordinate systems described by the abbreviations *LIA* and *LPS*. We see that the coronal plane corresponds to the third axes (A) in *LIA* and second axes (P) in *LPS*, and we have the opposite for the axial plane (I vs. S). Thus, these coordinate systems differ by the choice of a positive direction in the coronal and axial planes, in addition to their order.

Both coordinate systems describe voxel spaces, and we thus need to take into account any difference in voxel sizes. We can obtain voxel sizes (in millimeters) by further using `mri_info`:

```
$ cd $SUBJECTS_DIR/ernie/dti
$ mri_info tensor.nii.gz | grep voxel\ sizes
voxel sizes: 2.500000, 2.500000, 2.500000
```

```
$ cd $SUBJECTS_DIR/ernie/mri
$ mri_info orig.mgz | grep voxel\ sizes
voxel sizes: 1.000000, 1.000000, 1.000000
```

We observe that the voxel sizes differ, and therefore the transformation matrix needs to be scaled from 2.5 mm to 1.0 mm. Thus, the matrix transformation will have the form:

$$A = \begin{bmatrix} 0.4 & 0 & 0 \\ 0 & 0 & -0.4 \\ 0 & -0.4 & 0 \end{bmatrix}.$$

The vector b gives the difference between the origins of the two coordinate systems.

Note, however, that this affine transformation matrix is not quite realistic. First, it assumes that there is no rotational difference between the brains. Second, due to the lack of offset vector b as in (5.3), this transformation assumes that the origins have the same anatomical position. This is unlikely to be the case, since the magnetic resonance images differ in modality or occurrence (i.e. taken at different times). Therefore, to find the affine transformation matrix, we need to find the optimal overlap of the brain contour in the magnetic resonance images. This can be done manually, but it is preferable to do this using registration tools such as `bbregister`, which was used with `dt_recon` in Chapter 5.1.2. In our example, the affine transformation matrix can computed by taking the inverse of the augmented matrix found in `register.lta`[11] located in folder `$SUBJECTS_DIR/ernie/dti`. The augmented matrix is a combination of the matrix A and the vector b with the following structure:

$$\begin{bmatrix} A & b \\ 0 & 1 \end{bmatrix}$$

The approximated transformation matrix becomes

$$A = \begin{bmatrix} 0.4 & 0.0 & -0.1 \\ -0.1 & 0.0 & -0.4 \\ 0.0 & -0.4 & 0.0 \end{bmatrix},$$

and the translation vector

$$b = \begin{bmatrix} 9.0 \\ 106.4 \\ 7.7 \end{bmatrix}.$$

---

[11] This file was created by bbregister as part of the dt_recon command discussed in Section 5.1.2.

Chapter 6
Simulating anisotropic diffusion in heterogeneous brain regions

In this chapter, we return to our model problem (1.1) and bring together the tools and techniques introduced in Chapters 3 to 5. The computational domain will be determined from T1-weighted data and divided into gray and white matter subdomains, diffusion tensor imaging (DTI) data will be employed in the construction of the heterogeneous and anisotropic diffusion tensor, and specific sub regions, such as the hippocampus, will be selected to assess site-specific solute distribution governed by a diffusive process.

In practice, one should first address data and mesh resolution issues. For instance, raw DTI data can exhibit rough transitions, as well as noise. This is particularly true in the gray matter proximal to cerebrospinal fluid (e.g. Figures 5.2 and 5.4 in Chapter 5). Here, we assume that the DTI data have been suitably smoothened and denoised for use in simulations. In addition, we must ascertain a mesh resolution that will provide reliable estimates of the spread of different molecules, while avoiding the unnecessary computational costs associated with over-resolving the mesh.

6.1 Molecular diffusion in one dimension

To estimate a suitable spatial mesh resolution, time step, and time scale of the solute diffusion, it is useful to first consider equation (1.1) in one dimension for different molecules. Here, we consider the protein fragment amyloid-beta ($A\beta$) associated with neurodegenerative disease [33], the tracer gadobutrol used in glymphatic magnetic resonance imaging [54], and water. The effective diffusion

© The Author(s) 2022
K.-A. Mardal et al., *Mathematical Modeling of the Human Brain*,
Simula SpringerBriefs on Computing 10,
https://doi.org/10.1007/978-3-030-95136-8_6

coefficient D in brain tissue for each of these molecules is estimated to be 6.2×10^{-5} mm^2/s, 1.3×10^{-4} mm^2/s, and 1.1×10^{-3} mm^2/s, respectively [69, 67].

6.1.1 Analytical solution

In one dimension and over the domain $(0, \infty)$, the parabolic diffusion problem (1.1) with $u_0(x) = 0$, $u(0, t) = 1$, and $u(\infty, t) = 0$ allows for a simple analytic solution:

$$u(x, t) = \mathrm{erfc}(x/(2\sqrt{Dt})). \tag{6.1}$$

Figure 6.1 shows solutions of (6.1) zoomed in on the (left) first 2 mm of the domain, and the (middle) first 10 mm after 9 hours, and (right) the first 10 mm after 24 hours. It is evident that diffusion is a slow process: significant concentration changes occur within 2 mm of the boundary after 9 hours; however, 1 cm away, the heavier molecules, amyloid-beta and gadobutrol, still have concentrations near zero. The source code for generating these plots is available in `mri2fem/chp6/analytical_1D.py`.

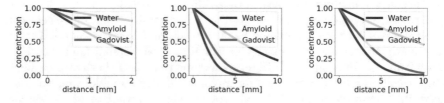

Fig. 6.1 Diffusion according to (6.1): concentration (arbitrary unit) versus distance from the source/left boundary after 9 hours (left and middle) and after 24 hours (right).

6.1.2 Numerical solution and handling numerical artifacts

Next, we discretize (1.1) using the finite element method (as described in Chapter 3). Note, however, that the sharp change in the boundary versus initial conditions for our model problem can lead to artificial oscillations in the numerical solution. Such oscillations often diminish with refinement; they can also be avoided through the use of monotonic or maximum principle preserving schemes. Another common method, which we consider here, for Galerkin finite element schemes is mass lumping (e.g. [41]). We provide FEniCS-based source code for the finite element solution of (1.1) with and without mass lumping in `mri2fem/chp6/diffusion_1D.py`. To use this script, see, for example:

```
$ cd mri2fem/chp6
$ python3 diffusion_1D.py --help
```

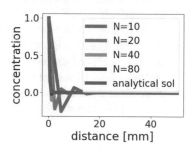

(a) $t = 30$ minutes, standard Galerkin

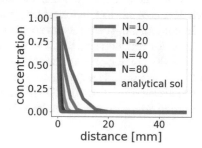

(b) $t = 30$ minutes, lumped mass matrix

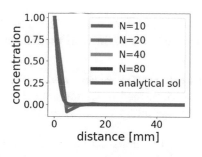

(c) $t = 9$ hours, standard Galerkin

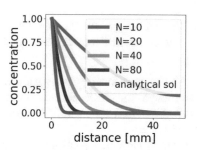

(d) $t = 9$ hours, lumped mass matrix

Fig. 6.2 Comparison of standard Galerkin (left) and mass-lumped (right) finite element schemes of the diffusion equation (1.1) in one dimension over $\Omega = (0, 50)$ mm at different times. The parameter N is the number of finite elements and the time step is 5 minutes.

The standard and mass-lumped finite element solutions are shown in Figure 6.2 at different times and with different time steps. Early on, for coarse resolutions ($N = 10$ or $N = 20$), with mesh size parameter $h = 50\,\text{mm}/N$, the standard approach yields considerable nonphysical oscillations, whereas the mass-lumped solution (right) produces significant numerical diffusion. However, in the longer-term context, the standard Galerkin scheme is clearly desirable: the former allows for a spatial resolution of $N = 10$ or $N = 20$, whereas the latter requires $N = 40$ or $N = 80$ to control the numerical diffusion. Essentially, the initial error from the short-term Gibbs phenomenon, that is, the discontinuous initial data, is no match for the long-term regularizing effect of the parabolic partial differential equation. Therefore, these early errors do not contribute much to the long-term numerical solution.

In conclusion, these results suggest that, if we are interested in long-term dynamics, a time step size of $\Delta t \approx 5$ minutes with a spatial resolution of $N = 40$ or $N = 80$, corresponding roughly to a quasi-uniform mesh cell diameter of 0.5 mm $\leq h \leq 1$ mm, is a good starting target for the standard Galerkin approach in our three dimensional (3D) discretization. The corresponding scheme with a lumped mass matrix does, however, significantly overestimate the diffusion at $N = 80$.

6.2 Anisotropic diffusion in 3D brain regions

In this section, we consider simulations of gadobutrol diffusion and compute the average concentrations in different brain regions. We begin with the following steps:

- We create a brain mesh with gray and white matter marked and ventricles removed and mark parcellation regions as described in Chapter 4.4.2.
- We filter and map our DTI data onto this geometry as described in Chapter 5.2.2.
- Using FEniCS, we implement a version of the diffusion simulation script presented in Chapter 3.3.3 allowing for anisotropic diffusion and the computation of integrals over labeled regions.

In the numerical simulation, we represent the DTI data in the form of a heterogeneous and anisotropic diffusion tensor field D. The FEniCS code for setting up the diffusion tensor field reads:

```
# read the DTI
```

```
T = TensorFunctionSpace(mesh, "DG", 0)
D = Function(T)
hdf.read(D, "/DTI")
```

We compute the average amount of tracer in a labeled region by integrating the concentration over the region and dividing by the region's volume as follows (with the regions labeled 17 and 1035 as examples):

```
    unit17 += [assemble(u*dx(17))/vol17]
    unit1035 += [assemble(u*dx(1035))/vol1035]
```

The precise commands run are included in mri2fem/chp6/all.sh, and the script mri2fem/chp6/chp6-diffusion-mritracer.py gives the complete FEniCS code.

6.2.1 Regional distribution of gadobutrol

We compute the average concentrations of gadobutrol diffusing in from the brain's surface in regions 17 (hippocampus), 1035 (insula gray matter), 3035 (insula white matter), 1028 (superior frontal gray matter), and 3028 (superior frontal white matter). The diffusivity of Gadobutrol is approximately twice that of amyloid-beta, and the estimated mesh size and time step of the previous section should therefore apply to this case as well. The resulting curves are shown in Figure 6.3, and the simulation results are shown in Figures 6.4–6.5. Note that, here, we consider the tracer distribution in certain regions as a function of time; the distribution therefore starts at a low value and increases with time as the solute diffuses throughout the brain. Clearly, the distribution of gadobutrol in the gray matter regions and hippocampus are affected much more than in the white matter regions. This result is expected since both the gray matter and hippocampus are closer to the cerebrospinal fluid where, in our simulation, the gadobutrol concentration is assumed to reside initially. It is also observed that the upper regions, that is, the superior frontal gray and white matter (1028 and 3028, respectively), experience faster gadobutrol deposition than the corresponding regions on the side of the brain.

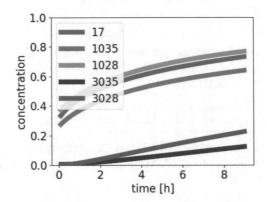

Fig. 6.3 Average concentration of gadobutrol (y-axis, arbitrary unit) versus time (x-axis, hours) in different brain regions: 17 (hippocampus), 1035 (insula gray matter), 3035 (insula white matter), 1028 (superior frontal gray matter), and 3028 (superior frontal white matter). Time step: 6 minutes, $N = 64$ brain mesh (cf. below).

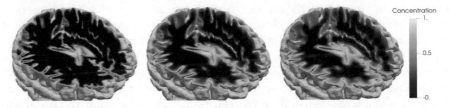

Fig. 6.4 The simulated distribution of gadobutrol, for a mesh with resolution parameter set to 32, after 0 hours (left), 5 hours (middle) and 9 hours (right).

6.2.2 Accuracy and convergence of computed quantities

A common question in numerical simulations is whether the computed solutions have converged. In this section, we therefore investigate the mesh convergence of the standard Galerkin and mass-lumped Galerkin approaches. More precisely, we consider a set of meshes, aiming to determine the accuracy of the numerical solution. In this example, we consider a roughly uniform refinement, but the mesh is not refined in place; rather, a sequence of meshes is first generated at different resolutions using the surface volume meshing toolkit

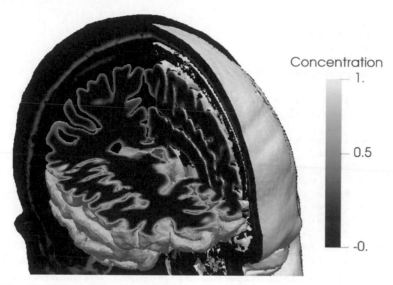

Fig. 6.5 Illustration of the simulated distribution of solute concentration in the brain within the cranium.

(SVM-Tk). In particular, we construct a sequence of quasi-uniform meshes, as follows (using `mri2fem/chp6/create_mesh_refinements.py`):

```python
import SVMTK as svmtk
import time

# Import surfaces, and merge lh/rh white surfaces
ventricles  = svmtk.Surface("surfaces/lh.ventricles.stl")
lhpial = svmtk.Surface("surfaces/lh.pial.stl")
rhpial = svmtk.Surface("surfaces/rh.pial.stl")
white = svmtk.Surface("surfaces/lh.white.stl")
rhwhite = svmtk.Surface("surfaces/rh.white.stl")
white.union(rhwhite)

surfaces = [lhpial, rhpial, white, ventricles]

# Create subdomain map
smap = svmtk.SubdomainMap()
smap.add("1000", 1)
smap.add("0100", 1)
smap.add("0110", 2)
smap.add("0010", 2)
smap.add("1010", 2)
```

```
smap.add("0111", 3)
smap.add("1011", 3)

# Create domain
domain = svmtk.Domain(surfaces, smap)

# Create meshes of increasing resolutions
Ns = [16, 32, 64, 128]
for N in Ns:
    print("Creating mesh for N=%d" % N)
    t0 = time.time()
    domain.create_mesh(N)
    domain.remove_subdomain([3])
    domain.save("brain_%d.mesh" % N)
    t1 = time.time()
    print("Done! That took %g sec" % (t1-t0))
```

After creating the meshes we mark the subdomains of interest and map the DTI data onto the mesh, before running the simulations. The following is a code snippet from `mri2fem/chp6/all.sh` that shows how the 16 mesh is created by the scripts described in the previous chapters:

```
# using the 16 mesh
# convert to h5
python3 ../chp4/convert_to_dolfin_mesh.py \
    --meshfile brain_16.mesh --hdf5file brain_16.h5

# mark subdomains
python3 ../chp4/add_parcellations.py \
    --in_hdf5 brain_16.h5 \
    --in_parc ../chp4/wmparc.mgz \
    --out_hdf5 brain_16_tags.h5 \
    --add 17 1028 1035 3028 3035

# add dti to the h5 file
python3 ../chp5/dti_data_to_mesh.py  \
    --dti ../chp5/clean-dti.mgz \
    --mesh brain_16_tags.h5 --label 1 0.4 0.6 \
    --out DTI_16.h5

# run simulation
python3 chp6-diffusion-mritracer.py --mesh DTI_16.h5 \
    --lumped lumped --annotation uniform16lumped
```

```
python3 chp6-diffusion-mritracer.py --mesh DTI_16.h5 \
    --lumped not --annotation uniform16notlumped
```

The average gadobutrol concentrations in the hippocampus over time for the sequence of meshes generated here are shown in Figure 6.6, without (left) and with (right) mass lumping. Clearly, the standard Galerkin approach (left) seems to yield more consistent results than the mass-lumped Galerkin scheme (right). However, even for the standard Galerkin scheme, whether the solutions are fully converged seems questionable at the highest resolution tested (around 15.5 million mesh cells). Recall that piecewise constants are used to represent the anisotropic diffusion tensor D. This DG construction requires about nine entries per cell, thus yielding approximately 140 million values for 15.5 million cells. Higher resolutions, such as those for piecewise linear or quadratic constructions, are not feasible on a personal computing device with only 32 gigabytes of RAM.

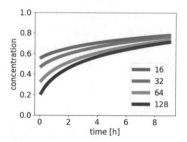 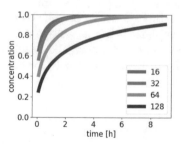

Fig. 6.6 Average gadobutrol concentration in the hippocampus (y-axis, arbitrary unit) versus time (x-axis, hours) for different mesh resolutions, $\Delta t = 6$ min. Quasi-uniform mesh sequence with $N = 16, 32, 64, 128$ generated by SVM-Tk. Standard Galerkin (left) versus mass-lumped Galerkin (right) discretizations.

To further assess the accuracy and convergence of the computed concentrations under mesh refinements, we therefore also consider adaptively refined meshes. In particular, we focus on the hippocampus and adaptively refine the meshes in this region, starting from the $N = 16$ brain mesh of the previous mesh sequence. Again, we plot the average gadobutrol concentrations in the hippocampus over time for a sequence of adaptively refined meshes (see Figure 6.7 with (right) and without (left) mass-lumping). Using this technique, we find the solutions between the first, second, and third adaptive refinements differ little for the standard scheme. However, mesh convergence for the mass-lumped Galerkin strategy remains unclear, even after four refinements to the hippocampal region.

Finally, we examine the mesh statistics (e.g. the number of vertices, cells and the range of mesh sizes) for the uniformly and adaptively refined meshes. Before doing so, we comment on the variables h, h_{max} and h_{min}. In scientific computing, h typically refers to a quantity defined on each tetrahedron T in the mesh and $h = h_T = \max\{x - y\}$ where x and y are any two points in T. We can then define the max and min values as $h_{max} = \max h_T$ and h_{min} is defined similarly; the T subscript on h_T is typically dropped and we have $h_{min} \leq h \leq h_{max}$.

Figure 6.6 suggests that, on quasi-uniform meshes, we reach mesh convergence around the refinement level denoted by '128', which, as shown in Table 6.1 (left), consists of about 15.5 million tetrahedrons. Figure 6.7 suggests that when we use adaptive refinement, mesh convergence is reached at refinement level '4' which consists of nearly 7.7 million tetrahedrons (Table 6.1 (right)). Thus, adaptive refinement has reduced the number of mesh tetrahedrons by half; a clear benefit. Using our one-dimension test case, in Chapter 6.1, we estimated that a value of $h_{min} \approx 0.5$ mm would be needed to reach mesh convergence. However, the results of Table 6.1 indicate that our estimate was off by a factor of about 3, for the quasi-uniform case, or 10, for the adaptive case.

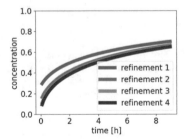 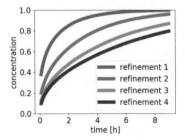

Fig. 6.7 Average gadobutrol concentration in the hippocampus (y-axis, arbitrary unit) versus time (x-axis, hours) for a sequence of adaptively refined meshes, $\Delta t = 6$ minutes. Standard Galerkin (left) versus mass-lumped Galerkin (right) discretizations.

Refinement	Vertices	Cells	h_{min}	h_{max}
16	94K	457K	0.97	11.4
32	194K	908K	0.46	5.7
64	567K	2.75M	0.26	2.9
128	2.8M	15.5M	0.14	1.45

Refinement	Vertices	Cells	h_{min}	h_{max}
1	99K	479K	0.64	11.4
2	123K	613K	0.30	11.4
3	275K	1.5M	0.14	11.4
4	1.3M	7.7M	0.07	11.4

Table 6.1 Mesh statistics (number of vertices, cells, and minimal and maximal cell sizes) for the (left) uniformly refined and (right) adaptively refined mesh sequences.

In summary, we have established that assessing the process of tracer distribution within the brain due to diffusion is a feasible, but somewhat computationally demanding task. In our case, we focused on the hippocampus and Gadobutrol enrichment and found that indeed a standard Galerkin procedure was sufficient given a locally refined mesh of a few hundred thousand cells. Quasi-uniform meshes on the other hand need several million cells before convergence. It is also worth noting that mass-lumping schemes, which are often prefered due to their monotonic properties that reduce non-physical oscillations in the short-term, can suffer from significant added numerical diffusivity and corresponding non-physical spread of tracer in long-term scenarios.

Chapter 7
Concluding remarks and outlook

Physics-based modeling of brain mechanics, informed by multi-modal imaging data, is an exciting research area at the frontier of science. Through this introductory and hands-on text, we have aimed to present a sufficient yet accessible amount of material to place the reader near the research front and to establish a solid foundation for further scientific investigations.

The MRI data to finite element pipeline presented here can be used for further simulations of solute transport within the brain, and importantly, provides a basis for more complex simulation scenarios. Our techniques for creating finite element meshes from T1 MR images can be equally applicable for numerical simulation of the brain as an elastic medium in the context of traumatic brain injuries, as a poroelastic medium for studying neurological disorders or stroke, or as an electrical medium for studying the propagation of epileptic seizures. Similarly, anisotropy data extracted from DTI can inform diffusion tensors (as here), or the permeability tensor in the context of brain fluid movement, the conductivity tensor in the context of brain electrophysiology, or possibly the compliance tensor(s) in the context of brain elasticity. While such physical models have not been considered here, the meshes and finite element simulation platform FEniCS extend readily to these contexts.

The brain does not exist in isolation, but is tightly coupled to its local environment, including the surrounding CSF, vasculature, membranes and spinal cord. The components presented here can be used as building blocks for computational modeling of the brain and its local environment, but geometrical, numerical, and computational challenges remain for this non-trivial multi-physics setting.

© The Author(s) 2022
K.-A. Mardal et al., *Mathematical Modeling of the Human Brain*,
Simula SpringerBriefs on Computing 10,
https://doi.org/10.1007/978-3-030-95136-8_7

In coming decades, we envisage that mathematics and numerical computations could play a crucial role in gaining new insights into the mechanisms and physiological processes of the human brain. Indeed, clinicians and experimentalists express a need for modeling and simulation as an alternative avenue of investigation to alleviate fundamental limitations in traditional techniques. A greater understanding of physiology and pathology could then ultimately pave the way for new diagnostics and treatments for a range of brain disorders and diseases — with immense scientific and societal impact.

References

1. *FEniCS Project*, 2020, https://fenicsproject.org.
2. *FreeSurfer*, 2020, https://surfer.nmr.mgh.harvard.edu/.
3. *FreeSurfer Wiki*, 2020, https://surfer.nmr.mgh.harvard.edu/fswiki.
4. *Paraview*, 2020, https://www.paraview.org.
5. N. J. ABBOTT, M. E. PIZZO, J. E. PRESTON, D. JANIGRO, AND R. G. THORNE, *The role of brain barriers in fluid movement in the CNS: is there a "glymphatic" system?*, Acta neuropathologica, 135 (2018), pp. 387–407.
6. J. AHRENS, B. GEVECI, AND C. LAW, *Paraview: An end-user tool for large data visualization*, The visualization handbook, 717 (2005).
7. R. ALDEA, R. O. WELLER, D. M. WILCOCK, R. O. CARARE, AND G. RICHARDSON, *Cerebrovascular smooth muscle cells as the drivers of intramural periarterial drainage of the brain*, Frontiers in aging neuroscience, 11 (2019), p. 1.
8. A. L. ALEXANDER, J. E. LEE, M. LAZAR, AND A. S. FIELD, *Diffusion tensor imaging of the brain*, Neurotherapeutics, 4 (2007), pp. 316–329.
9. M. ALNÆS, J. BLECHTA, J. HAKE, A. JOHANSSON, B. KEHLET, A. LOGG, C. RICHARDSON, J. RING, M. E. ROGNES, AND G. N. WELLS, *The FEniCS project version 1.5*, Archive of Numerical Software, 3 (2015).
10. N. ALPERIN AND A. M. BAGCI, *Spaceflight-induced visual impairment and globe deformations in astronauts are linked to orbital cerebrospinal fluid volume increase*, in Intracranial Pressure & Neuromonitoring XVI, Springer, 2018, pp. 215–219.
11. K. ARCHIE, *Dicombrowser*, 2020, https://nrg.wustl.edu/software/dicom-browser/.
12. A. BERGER, *Magnetic resonance imaging*, BMJ, 324 (2002), p. 35, https://doi.org/10.1136/bmj.324.7328.35.
13. W. F. BORON AND E. L. BOULPAEP, *Medical Physiology*, Elsevier Health Sciences, 2016.
14. M. BRETT ET AL., *nipy/nibabel: 3.2.1*, Nov. 2020, https://doi.org/10.5281/zenodo.4295521.
15. S. BUDDAY, T. C. OVAERT, G. A. HOLZAPFEL, P. STEINMANN, AND E. KUHL, *Fifty shades of brain: a review on the mechanical testing and modeling of brain tissue*, Archives of Computational Methods in Engineering, (2019), pp. 1–44.

© The Author(s) 2022
K.-A. Mardal et al., *Mathematical Modeling of the Human Brain*,
Simula SpringerBriefs on Computing 10,
https://doi.org/10.1007/978-3-030-95136-8

16. D. CHOU, J. C. VARDAKIS, L. GUO, B. J. TULLY, AND Y. VENTIKOS, *A fully dynamic multi-compartmental poroelastic system: Application to aqueductal stenosis*, Journal of biomechanics, 49 (2016), pp. 2306–2312.

17. M. CROCI, V. VINJE, AND M. E. ROGNES, *Uncertainty quantification of parenchymal tracer distribution using random diffusion and convective velocity fields*, Fluids and Barriers of the CNS, 16 (2019), p. 32.

18. A. M. DALE, B. FISCHL, AND M. I. SERENO, *Cortical surface-based analysis: I. segmentation and surface reconstruction*, Neuroimage, 9 (1999), pp. 179–194.

19. C. DAVERSIN-CATTY, V. VINJE, K.-A. MARDAL, AND M. E. ROGNES, *The mechanisms behind perivascular fluid flow*, Plos one, 15 (2020), p. e0244442.

20. A. K. DIEM, M. MACGREGOR SHARP, M. GATHERER, N. W. BRESSLOFF, R. O. CARARE, AND G. RICHARDSON, *Arterial pulsations cannot drive intramural periarterial drainage: significance for Aβ drainage*, Frontiers in neuroscience, 11 (2017), p. 475.

21. A. FABRI, G.-J. GIEZEMAN, L. KETTNER, S. SCHIRRA, AND S. SCHÖNHERR, *On the design of CGAL, a computational geometry algorithms library*, Software: Practice and Experience, 30 (2000), pp. 1167–1202.

22. S. FORNARI, A. SCHÄFER, M. JUCKER, A. GORIELY, AND E. KUHL, *Prion-like spreading of alzheimer's disease within the brain's connectome*, Journal of the Royal Society Interface, 16 (2019), p. 20190356.

23. C. GEUZAINE AND J.-F. REMACLE, *Gmsh: A 3-d finite element mesh generator with built-in pre-and post-processing facilities*, International journal for numerical methods in engineering, 79 (2009), pp. 1309–1331.

24. M. S. GOCKENBACH, *Understanding and implementing the finite element method*, vol. 97, SIAM, 2006.

25. A. GORIELY, M. G. GEERS, G. A. HOLZAPFEL, J. JAYAMOHAN, A. JÉRUSALEM, S. SIVALOGANATHAN, W. SQUIER, J. A. VAN DOMMELEN, S. WATERS, AND E. KUHL, *Mechanics of the brain: perspectives, challenges, and opportunities*, Biomechanics and modeling in mechanobiology, 14 (2015), pp. 931–965.

26. H. GRAY, *Gray's anatomy: with original illustrations by Henry Carter*, Arcturus Publishing, 2009.

27. R. GRECH, T. CASSAR, J. MUSCAT, K. P. CAMILLERI, S. G. FABRI, M. ZERVAKIS, P. XANTHOPOULOS, V. SAKKALIS, AND B. VANRUMSTE, *Review on solving the inverse problem in EEG source analysis*, Journal of neuroengineering and rehabilitation, 5 (2008), p. 25.

28. A. GREENBAUM, *Iterative Methods for Solving Linear Systems*, Society for Industrial and Applied Mathematics, 1997, https://doi.org/10.1137/1.9781611970937.

29. E. M. HAACKE, R. W. BROWN, M. R. THOMPSON, R. VENKATESAN, M. THOMPHSON, M. VENKATESAN, M. HAACKE, W. BROWN, AND M. THOMPSON, *Magnetic resonance imaging: physical principles and sequence design*, (1999).

30. P. T. HAGA, G. PIZZICHELLI, M. MORTENSEN, M. KUCHTA, S. H. PAHLAVIAN, E. SINIBALDI, B. A. MARTIN, AND K.-A. MARDAL, *A numerical investigation of intrathecal isobaric drug dispersion within the cervical subarachnoid space*, PloS one, 12 (2017), p. e0173680.

31. K. E. HOLTER, B. KEHLET, A. DEVOR, T. J. SEJNOWSKI, A. M. DALE, S. W. OMHOLT, O. P. OTTERSEN, E. A. NAGELHUS, K.-A. MARDAL, AND K. H. PETTERSEN, *Interstitial solute transport in 3D reconstructed neuropil occurs by diffusion rather than bulk flow*, Proceedings of the National Academy of Sciences, 114 (2017), pp. 9894–9899.

32. J. D. Hunter, *Matplotlib: A 2D graphics environment*, Computing in science & engineering, 9 (2007), pp. 90–95.
33. J. J. Iliff, M. Wang, Y. Liao, B. A. Plogg, W. Peng, G. A. Gundersen, H. Benveniste, G. E. Vates, R. Deane, S. A. Goldman, et al., *A paravascular pathway facilitates CSF flow through the brain parenchyma and the clearance of interstitial solutes, including amyloid β*, Science translational medicine, 4 (2012), pp. 147ra111–147ra111.
34. J. J. Iliff, M. Wang, D. M. Zeppenfeld, A. Venkataraman, B. A. Plog, Y. Liao, R. Deane, and M. Nedergaard, *Cerebral arterial pulsation drives paravascular CSF–interstitial fluid exchange in the murine brain*, Journal of Neuroscience, 33 (2013), pp. 18190–18199.
35. M. Jenkinson, C. F. Beckmann, T. E. Behrens, M. W. Woolrich, and S. M. Smith, *FSL*, Neuroimage, 62 (2012), pp. 782–790.
36. N. A. Jessen, A. S. F. Munk, I. Lundgaard, and M. Nedergaard, *The glymphatic system: a beginner's guide*, Neurochemical research, 40 (2015), pp. 2583–2599.
37. B. Jeurissen, M. Descoteaux, S. Mori, and A. Leemans, *Diffusion MRI fiber tractography of the brain*, NMR Biomed., 32 (2017), p. e3785, https://doi.org/doi.org/10.1002/nbm.3785.
38. P. Kevrekidis, T. B. Thompson, and A. Goriely, *Anisotropic diffusion and traveling waves of toxic proteins in neurodegenerative diseases*, Physics Letters A, 384 (2020), p. 126935.
39. G. Kindlmann, R. S. J. Estepar, M. Niethammer, S. Haker, and C.-F. Westin, *Geodesic-loxodromes for diffusion tensor interpolation and difference measurement*, in International Conference on Medical Image Computing and Computer-Assisted Intervention, Springer, 2007, pp. 1–9.
40. P. Kochunov, D. C. Glahn, J. Lancaster, P. M. Thompson, V. Kochunov, B. Rogers, P. Fox, J. Blangero, and D. Williamson, *Fractional anisotropy of cerebral white matter and thickness of cortical gray matter across the lifespan*, Neuroimage, 58 (2011), pp. 41–49.
41. H. P. Langtangen and A. Logg, *Solving PDEs in Python: The FEniCS Tutorial I*, vol. 1, Springer, 2016.
42. H. P. Langtangen and K.-A. Mardal, *Introduction to numerical methods for variational problems*, vol. 21, Springer Nature, 2019.
43. J. J. Lee, E. Piersanti, K.-A. Mardal, and M. E. Rognes, *A mixed finite element method for nearly incompressible multiple-network poroelasticity*, SIAM Journal on Scientific Computing, 41 (2019), pp. A722–A747.
44. A. Logg, K.-A. Mardal, and G. Wells, *Automated solution of differential equations by the finite element method: The FEniCS book*, vol. 84, Springer Science & Business Media, 2012.
45. K.-A. Mardal, M. E. Rognes, T. B. Thompson, and L. M. Valnes, *mri2fem data set*, 2021, https://doi.org/10.5281/zenodo.4386986.
46. K.-A. Mardal, M. E. Rognes, T. B. Thompson, and L. M. Valnes, *Software for Mathematical modeling of the human brain – from magnetic resonance images to finite element simulation*, 2021, https://doi.org/10.5281/zenodo.4386998.
47. A. Meurer, C. P. Smith, M. Paprocki, O. Čertík, S. B. Kirpichev, M. Rocklin, A. Kumar, S. Ivanov, J. K. Moore, S. Singh, et al., *SymPy: symbolic computing in Python*, PeerJ Computer Science, 3 (2017), p. e103.

48. P. MILDENBERGER, M. EICHELBERG, AND E. MARTIN, *Introduction to the DICOM standard*, European radiology, 12 (2002), pp. 920–927.

49. O. NAGGARA, C. OPPENHEIM, D. RIEU, N. RAOUX, S. RODRIGO, G. DALLA BARBA, AND J.-F. MEDER, *Diffusion tensor imaging in early Alzheimer's disease*, Psychiatry Research: Neuroimaging, 146 (2006), pp. 243–249.

50. S. J. PAYNE, *Cerebral Blood Flow and Metabolism: A Quantitative Approach*, World Scientific, 2017.

51. G. PIZZICHELLI, B. KEHLET, Ø. EVJU, B. MARTIN, M. ROGNES, K. MARDAL, AND E. SINIBALDI, *Numerical study of intrathecal drug delivery to a permeable spinal cord: effect of catheter position and angle*, Computer Methods in Biomechanics and Biomedical Engineering, 20 (2017), pp. 1599–1608.

52. R. A. POOLEY, *Fundamental physics of mr imaging*, Radiographics, 25 (2005), pp. 1087–1099.

53. L. RAY, J. J. ILIFF, AND J. J. HEYS, *Analysis of convective and diffusive transport in the brain interstitium*, Fluids and Barriers of the CNS, 16 (2019), p. 6.

54. G. RINGSTAD, L. M. VALNES, A. M. DALE, A. H. PRIPP, S.-A. S. VATNEHOL, K. E. EMBLEM, K.-A. MARDAL, AND P. K. EIDE, *Brain-wide glymphatic enhancement and clearance in humans assessed with MRI*, JCI insight, 3 (2018).

55. L. ROSCOE ET AL., *Stereolithography interface specification*, America-3D Systems Inc, 27 (1988), p. 10.

56. N. SCHLÖMER, G. MCBAIN, T. LI, V. M. FERRÁNDIZ, EOLIANOE, L. DALCIN, K. LUU, NILSWAGNER, A. GUPTA, S. MÜLLER, L. SCHWARZ, J. BLECHTA, C. COUTINHO, D. BEURLE, B. SHRIMALI, A. CERVONE, NATE, U. MISHRA, T. HEISTER, T. LANGLOIS, S. PEAK, S. SHARMA, M. BUSSONNIER, LGIRALDI, G. JACQUENOT, G. A. VAILLANT, C. WILSON, A. U. GUDCHENKO, AND A. CROUCHER, *nschloe/meshio 3.2.14*, Nov. 2019, https://doi.org/10.5281/zenodo.3548723.

57. W. B. SCOVILLE AND B. MILNER, *Loss of recent memory after bilateral hippocampal lesions*, Journal of neurology, neurosurgery, and psychiatry, 20 (1957), p. 11.

58. M. K. SHARP, R. O. CARARE, AND B. A. MARTIN, *Dispersion in porous media in oscillatory flow between flat plates: applications to intrathecal, periarterial and paraarterial solute transport in the central nervous system*, Fluids and Barriers of the CNS, 16 (2019), p. 13.

59. A. J. SMITH, X. YAO, J. A. DIX, B.-J. JIN, AND A. S. VERKMAN, *Test of the "glymphatic" hypothesis demonstrates diffusive and aquaporin-4-independent solute transport in rodent brain parenchyma*, Elife, 6 (2017), p. e27679.

60. J. SOARES, P. MARQUES, V. ALVES, AND N. SOUSA, *A hitchhiker's guide to diffusion tensor imaging*, Frontiers in neuroscience, 7 (2013), p. 31.

61. L. R. SQUIRE, *The legacy of patient HM for neuroscience*, Neuron, 61 (2009), pp. 6–9.

62. K. STÜBEN, *A review of algebraic multigrid*, J. Comp. Appl. Math, 128 (2001), pp. 281–309.

63. E. SYKOVÁ AND C. NICHOLSON, *Diffusion in brain extracellular space*, Physiological reviews, 88 (2008), pp. 1277–1340.

64. G. TAUBIN, *Curve and surface smoothing without shrinkage*, in Proceedings of IEEE international conference on computer vision, IEEE, 1995, pp. 852–857.

65. A. THOMAS AND A. BANERJEE, *The History of Radiology (Oxford Medical Histories)*, Oxford University Press, 2013.

66. A. TVEITO AND R. WINTHER, *Introduction to partial differential equations: a computational approach*, vol. 29, Springer Science & Business Media, 2004.

67. L. M. VALNES, S. K. MITUSCH, G. RINGSTAD, P. K. EIDE, S. W. FUNKE, AND K.-A. MARDAL, *Apparent diffusion coefficient estimates based on 24 hours tracer movement support glymphatic transport in human cerebral cortex*, Scientific Reports, 10 (2020), pp. 1–12.

68. I. VOLDSBEKK, I. I. MAXIMOV, N. ZAK, D. ROELFS, O. GEIER, P. DUE-TØNNESSEN, T. ELVSÅSHAGEN, M. STRØMSTAD, A. BJØRNERUD, AND I. GROOTE, *Evidence for wakefulness-related changes to extracellular space in human brain white matter from diffusion-weighted MRI*, NeuroImage, (2020), p. 116682.

69. J. WATERS, *The concentration of soluble extracellular amyloid-β protein in acute brain slices from crnd8 mice*, PLoS One, 5 (2010), p. e15709.

70. L. XIE, H. KANG, Q. XU, M. J. CHEN, Y. LIAO, M. THIYAGARAJAN, J. O'DONNELL, D. J. CHRISTENSEN, C. NICHOLSON, J. J. ILIFF, ET AL., *Sleep drives metabolite clearance from the adult brain*, science, 342 (2013), pp. 373–377.

Index

© The Author(s) 2022   117
K.-A. Mardal et al., *Mathematical Modeling of the Human Brain*,
Simula SpringerBriefs on Computing 10,
https://doi.org/10.1007/978-3-030-95136-8

Printed in the United States
by Baker & Taylor Publisher Services